REMOTE LABORATORIES

Empowering STEM Education with Technology

REMOTE LABORATORIES

Empowering STEM Education with Technology

Javier García-Zubía

University of Deusto, Spain

In collaboration with Luis Rodríguez-Gil

World Scientific

NEW JERSEY • LONDON • SINGAPORE • BEIJING • SHANGHAI • HONG KONG • TAIPEI • CHENNAI • TOKYO

Published by

World Scientific Publishing Europe Ltd.
57 Shelton Street, Covent Garden, London WC2H 9HE
Head office: 5 Toh Tuck Link, Singapore 596224
USA office: 27 Warren Street, Suite 401-402, Hackensack, NJ 07601

Library of Congress Cataloging-in-Publication Data
Names: García-Zubía, Javier, author.
Title: Remote laboratories: empowering STEM education with technology / Javier García-Zubía, University of Deusto, Spain ; with the collaboration of Luis Rodríguez-Gil.
Description: London : World Scientific, 2021. | Includes bibliographical references and index.
Identifiers: LCCN 2020038666 (print) | LCCN 2020038667 (ebook) | ISBN 9781786349422 (hardcover) | ISBN 9781786349439 (ebook) | ISBN 9781786349446 (ebook other)
Subjects: LCSH: Science--Study and teaching. | Internet in education.
Classification: LCC Q181 .G244 2021 (print) | LCC Q181 (ebook) | DDC 507.8--dc23
LC record available at https://lccn.loc.gov/2020038666
LC ebook record available at https://lccn.loc.gov/2020038667

British Library Cataloguing-in-Publication Data
A catalogue record for this book is available from the British Library.

For any available supplementary material, please visit
https://www.worldscientific.com/worldscibooks/10.1142/Q0277#t=suppl

Desk Editors: Michael Beale/Shi Ying Koe/Aanand Jayaraman

Typeset by Stallion Press
Email: enquiries@stallionpress.com

Prologue

September 22, 2005. A lecture room at the University of Catania, Italy, is being used for hosting a technical session of the 10th IEEE International Conference on Emerging Technologies and Factory Automation (ETFA'05). Inside the room, two researchers (Gustavo Alves and Javier García Zubía) who had never met before are expected to present their work about the same topic: remote laboratories. Gustavo's presentation focuses on preliminary results of the RexNet project. His PowerPoint presentation contains seven slides for a 10 minutes' presentation. Results from most recent conversations among the project participants are still fresh on his mind and missing from the prepared slides. Luckily, there is a whiteboard and some marker pens in the room. He takes one pen and writes the following equation on the whiteboard, referring to the added value of remote experimentation to education:

Educational value ↑ — { Groupwork with foreign students; Share good educational practices }
\- Costs ↓ — { Share experiments; Reuse software }
Added value ↑

q.e.d.

FORTUNE[1] and HARD WORK:

The fact Gustavo had no slide with this simple equation combined with his will to share information led him to write the equation on the whiteboard. This unusual gesture, in the context of a formal presentation within a technical conference session, made an impact in Javier's mind. The fact Javier and Gustavo were not supposed to present their works at the same session (one of them had asked to change the presentation slot in order to catch a flight later that same day) combined with the coincidence of both meeting each other at the Catania International Airport, where they continued to discuss about remote labs and remote experimentation, were a fortunate stroke of serendipity. **Hard work did the rest.**

Gustavo, as coordinator of the RexNet project, asks for an amendment to include the University of Deusto in the project consortium. This enables Javier and his research group to join the project and take part in the next project general meetings, held in Porto and Santiago de Chile, in June and December, the following year, respectively. In the meantime, i.e., November 16–17, 2006, Javier organizes an International Meeting on Professional Remote Laboratories, inviting Gustavo, Ingvar Gustavsson, Luis Gomes, Domenico Ponta, Cornel Samoila, Andreas Pester, and Dieter Mueller, among other researchers, to take part and join as authors in the first book about remote laboratories, jointly edited by him and Luis Gomes[2].

Shortly after, i.e., in 2008, Javier installs at the University of Deusto the first external node of the Virtual Instruments Systems in Reality (VISIR) remote lab, developed by Ingvar Gustavsson at the Blekinge Institute of Technology (BTH). During this period, Unai Hernández Jayo visits BTH to do part of his PhD work, supervised by Javier, in what may be seen as a good implementation example of the previously shown equation.

[1] *Often capitalized: A hypothetical force or personified power that unpredictably determines events and issues favorably or unfavorably.* From https://www.merriam-webster.com/dictionary/fortune, accessed July 22, 2020.

[2] *Advances on Remote Laboratories and e-Learning Experiences.* Editors: Luís Gomes (Polytechnic of Porto) and Javier García-Zubía (Universidad de Deusto). Universidad de Deusto, Bilbao, 2007. Chapter 14, p. 309 ISBN 978-84-9830-662-0.

Later, Javier will edit two more books about remote labs, one in 2011 and another in 2013, respectively[3,4]. Again, sharing good educational practices emerges as a motivation for these two books.

Covers of the three books published by the University of Deusto on the subject of remote experimentation.

During this period, Javier organizes the 9th edition of the most prestigious conference in the area of remote labs, the Remote Engineering and Virtual Instrumentation (REV) conference, in Bilbao, in June 2012. This is a unique opportunity to gather and interact with the worldwide community working in this area. Later on, in 2014, Gustavo organizes the 11th REV conference edition in Porto, and invites other distinguished researchers, like Lyle Feisel[5,6] to know more about the community working on remote labs. Again, Gustavo and Javier are able to intensively learn from their peers.

---

[3] *Using Remote Labs in Education: Two Little Ducks in Remote Experimentation.* Editors: Javier García-Zubía (Universidad de Deusto) and Gustavo R. Alves (Polytechnic of Porto). Universidad de Deusto, Bilbao, 2011. Chapter 22, p. 465 ISBN 978-84-9830-335-3.

[4] *IT Innovative Practices in Secondary Schools: Remote Experiments.* Editors: Olga Dziabenko and Javier García-Zubía (Universidad de Deusto). Universidad de Deusto, Bilbao, 2011. Chapter 13, p. 347 ISBN 978-84-15772-01-9.

[5] Lyle Feisel is a co-author of one of the most cited papers about the importance of laboratory work in engineering education (see footnote 6). According to Google Scholar, his paper has more than 1400 citations https://scholar.google.pt/citations?user=_0i-9koAAAAJ.

[6] Feisel, L. D. and Rosa, A. J. (2005). The role of the laboratory in undergraduate engineering education. *Journal of Engineering Education*, 94: 121–130. doi:10.1002/j.2168-9830.2005.tb00833.x.

First photo: Carisa Bohus, Gustavo Alves and WebLab-Deusto researchers during the REV conference in Bilbao.

Second photo: Lyle Feisel, Ingvar Gustavsson, Gustavo Alves and Javier García-Zubía during REV conference in Porto.

From 2012 onwards, Javier engages himself in a number of research projects related to remote experimentation, e.g., Go-Lab, ICo-op, NeReLa, VISIR+, PILAR, and Next-Lab, among others[7], building a remarkable

[7]https://weblab.deusto.es/website/projects.html.

experience in the use of remote labs for enhancing practical skills in engineering education.

The present book is a result of countless hours of hard work, pleasant conversations with his peers and, above all, an everyday reflection on the subject while walking from his home to the University of Deusto or taking his dog on a walk near his home. As a colleague and a friend, I had the pleasure to witness his career path. As a researcher, I had the unique privilege of reading this book in its early stages of development, and, in particular 'The Ten Commandments of Remote Experimentation', which perfectly summarise Javier's vision about the use of remote labs in engineering education. There could be no better summary for such a book.

Gustavo R. Alves
Polytechnic of Porto — School of Engineering

Foreword

Remote laboratories have always been a special field in learning tools. They have been studied in depth in terms of pedagogical outcomes, technological aspects, potential impact, overlaps (hybrid laboratories) and, especially, the many different remote laboratories. However, it is one of the few fields in learning tools that has no case where someone reaches a truly large number of users. Successful simulations, learning management systems, plagiarism tools, MOOCs, Learning Analytics tools and many other learning tools have millions of users, while most remote labs are typically used by lab creators, their colleagues, and their friends. There are few exceptions wherein many people use the same technology (such as iLab, or WebLab-Deusto: but not the same laboratory) or the same laboratory (such as VISIR: purchased by many universities which use different — and incompatible — configurations of the laboratory), but I do not think that any of these initiatives has reached one million users.

Does it mean that it has to be this way, or that it will be this way? Definitely not! Once a laboratory reaches a certain technology readiness to become a reliable product, things change fast. For this to happen, the laboratory must first be pedagogically sound and, second, technologically reliable. The first one is obvious: as in every learning tool, if the laboratory's pedagogical outcomes are not sound, the laboratory is a 'technological toy'. The role of technology is in exclusively delivering this pedagogical experience to the user — not more, and not less. If, again, as in every learning tool, technology creates user experience problems (complex user interface, web browser or network requirements, failing to integrate with the rest of the pedagogical tools), it will also have a lower impact. However, technology in remote labs has another aspect that is crucially different from other learning tools: it relies on a piece of hardware that in most cases is not replicated across various locations, and very often not even replicated in one location. This leads to critical uncertainties: 'Will the hardware be available tomorrow at 10 am, when I use it, or will there be someone else using it? What happens if the hardware is not working?' Remote labs have a lot to offer: most users in successful remote labs are extremely engaged and see the positive difference compared to other tools. However, in order to go from the current state of remote labs to the state where other learning tools are, it is necessary to address this gap, tackling these uncertainties.

This is the reason why LabsLand was created. After a decade working in the field of remote laboratories, some University of Deusto researchers (including the authors Javier García-Zubía, Luis Rodríguez-Gil, and myself among others) decided to build a company that provided product-level remote laboratories to schools and higher education. To do this, LabsLand relies on agreements with many universities in different continents. At times, LabsLand relies on its own technologies with different approaches to tackle the large load of users. For example, in September 2020 Oxford students were using Arduino robot laboratories, and when they were programming code they did not know if they were going to be using the robots in Spain, Colombia, South Africa or Costa Rica — and this is irrelevant from a pedagogical experience. What they needed to know was that a robot was going to be available for them within a short or no time. Even if a robot was broken, students would not notice, as LabsLand automatically checks it and removes it from the pool of available robots in a transparent way. And the same process has happened in the over 100,000 laboratory sessions by thousands of students so far in 2020.

These technologies, tools and paradigms are available for researchers in the field of remote laboratories to support them to move their remote laboratories to the next level. But first, they have to reach the first level, which is creating the laboratories focused on the pedagogical aspects and tackling the basic technical aspects. To this end, the best way is to start with this magnificent book, based on the long-term experience of its author, Javier García-Zubía, leader of the WebLab-Deusto research group.

Finally, to support readers during the use of this book, LabsLand provides in the following link with access to many of the remote laboratories of this book for individuals or for institutions (schools, colleges and universities) at a discounted price: https://labsland.com/s/empowering-stem-education.

Pablo Orduña, PhD
CEO, Co-founder of LabsLand

Preface

This book, titled *Remote Laboratories: Empowering STEM Education with Technology*, deals with the subject of remote experimentation and aims to appeal to a number of readers. Its main objective is not to appeal only to engineers, by virtue of technological analysis, but to attract and interest potential users of remote experimentation in the classroom.

A remote experiment is something real that takes place in front of us in a digital rather than physical realm. The internet brings us the experiment from wherever it happens to be, and presents it to us on an interface on a computer, a tablet, a smartphone, etc.

In essence, remote experimentation democratises access to science, not just for any student, but for anybody with a connection to the internet. The book includes examples for primary and secondary pupils, but also for sixth form, university or vocational training students: from Archimedes' Principle to a programmable robot, from the effect of caffeine on planarians to a radioactivity experiment.

The book seeks, by means of rigorous, academic discourse, to provide information and arguments for the use of remote experimentation in the classroom. What happens when a remote experiment is used in the classroom? How do teachers and students benefit? What are the challenges posed by remote experimentation? What are the risks? What is the cost? Who needs to be persuaded to introduce them into the classroom? These questions are central to the book. Clearly, when a teacher closes the classroom door, only he or she knows what goes on inside, but it is also clear that in order for changes to be consistent and effective, it is important that other involved agents offer their support: school directors, educational innovation centres, policymakers, teacher training centres, blogs and educational forums, etc. The educational ecosystem is rich and open, and all its members must participate.

The benefits of remote experimentation are all the more evident in the light of the COVID-19 experience: students and teachers all went home and their interaction became digital. Wouldn't the possibility of conducting scientific experiments without going to school have been the ideal solution? What happened in universities and their laboratories?

The book is structured in a way that makes it interesting for different readers. The electronics engineer will find technical problems and their solutions, while the educator can skip these issues and focus on the educational impact of the remote laboratory.

The book will be of the utmost interest to students in the Faculty of Education, since they are responsible for innovating and introducing emerging trends into the classroom. These are the readers who can enhance the book with their experience and criticism.

The book is divided into four parts. The first part addresses general and fundamental issues of remote experimentation, such as the advantages and disadvantages of a remote laboratory, and its characteristics. A careful reading of this section will reveal to the reader the potential nuances of remote experimentation.

The second part describes examples of remote experimentation and how to use them in the classroom. This section, in which psychological and pedagogical considerations establish the rhythm of the book, is directed towards readers and end users.

In the third part, the focus shifts from teachers and users to designers. In this case, the discourse is distinctly technological, and the uninterested reader can skip this section and still benefit from the rest of the book. This section was written in collaboration with Luis Rodríguez-Gil, thanks in part to the EU ERASMUS+ project 'e-LIVES — e-Learning InnoVative Engineering Solutions', (ref. 585938-EPP-1-2017-1-FR-EPPKA2-CBHE).

The fourth and final part offers the investigative reader elements of interest for PhD or Masters students. Innovation is a key element in this section, so there is explanation of the use of virtual reality in remote experiments, for example, or analysis of the usability and interfaces of a remote laboratory. Finally, aspects are considered that could be regarded as marginal, but which may interest the curious reader.

The book poses questions and usually provides answers, but also leaves discussions open so that readers, teachers and researchers can debate and answer on the basis of their own studies.

Finally, a key element of this book is the fact that readers have privileged access to the remote experiments conducted by the Labsland company.

Labsland is currently the only company that approaches remote experimentation as a professional educational service. This company was created five years ago as a spin-off from Deusto University, and all the founders belong to the university, including the author of this book.

Before concluding this preface, I would like to express my gratitude to Pablo Orduña. He was contacted by WSP and asked to write this book, and invited me to join him as co-author. Pablo Orduña is currently CEO of LabsLand, and this commitment has made it impossible for him to write and revise this volume. He placed his trust in me and encouraged me to assume responsibility for the book and the subsequent promotion of remote experimentation. I hope I have accomplished that mission. My sincerest thanks for the opportunity.

I hope that the book appeals to all readers and contributes to a positive evaluation of the use of remote experimentation in the scientific-technological training of students. Any contributions, criticism or suggestions will be welcome at zubia@deusto.es and will help the book continue to evolve.

About the Author

Javier García-Zubía is a Professor at the University of Deusto (Spain). He is the leader of the WebLab-Deusto research group in remote laboratories. His research is focused on remote experimentation: design, implementation and evaluation. He is author and co-author of dozens of works published in journals and in international conferences, and he has been involved in different international projects such as FP7, H2020, Erasmus+, NSF, etc. He has edited three books on remote experimentation and has organized many workshops and courses in this field. He has received several recognitions and awards.

Contents

Part

1

Remote Laboratories: General Concepts

Chapter 1

The Definition of Remote Experimentation

This first chapter describes what a remote experiment is and its main advantages and disadvantages. This, along with its historical evolution, creates a frame of reference for the rest of the book.

1.1 What Is a Remote Laboratory?

In an experiment, the student observes a physical phenomenon, changes process input variables and measures the behaviour of output variables. In Max Planck's words (https://en.wikiquote.org/wiki/Talk:Max_Planck):

> *'An experiment is a question which science poses to Nature, and a measurement is the recording of Nature's answer.'*

A remote experiment is an experiment mediated by the Internet. Basically, our hands and eyes will be the mouse and the webcam, all computer-supervised, and although the experiment is in a different location, its requirements and potential are the same. According to Ma and Nickerson (2006):

> ***'Remote labs are characterized by mediated reality. Similar to hands-on labs, they require space and devices. What makes them different from real labs is the distance between the experiment and the experimenter. In real labs, the equipment might be mediated through computer control, but collocated. By contrast, in remote labs experimenters obtain data by controlling geographically detached equipment. In other words, reality in remote labs is mediated by distance.'***

All our actions and all the results of the experiment will travel via the Internet. This is a clear example of Technology-Enhanced Learning (TEL), and whenever the student plays an active and relevant role in its use, it is more appropriate to speak of Technology-Enhanced Active Learning (TEAL). Both tendencies fall within what Ramo (1957) foresaw over 60 years ago:

> ***'There is probably a new profession known as teaching engineer, that kind of engineering which is concerned with the educational process and with the design of the machines, as well as the design of the material.'***

In a quality remote experiment, our global experience should be at least equivalent to the on-site experiment, if not superior.

Max Planck's aforementioned sentence, in the words of Ingvar Gustavsson — designer of the VISIR remote laboratory and Maffioli Award of Excellence for developing teaching and learning in engineering education — becomes:

> ***'Experimenting could be compared to a conversation with nature. The experimenter asks and nature answers. The tricky thing is formulating a useful question and above all interpreting the answer. The only way to learn the language of nature is performing many experiments in laboratories that can be hands-on or remote.'***

We can see that, for Ingvar Gustavsson, all that changes is that the laboratory can be in front of us or be remote, in which case it will generally be necessary to control it via the Internet. Conceptually, there is no difference between a hands-on and a remote laboratory, although technology will establish certain differences.

Before continuing, it should be noted that there are various equivalent terms within remote experimentation. The most suitable name is *Remote Labs* and *Remote Experiments*, but there are also authors who use the term *Virtual Labs*, which should be employed only for those laboratories that use simulators fundamentally based on mathematical models. The terms *WebLab* and *iLab* are also used, both introduced by MIT, where 'web' and 'i' highlight the importance of the Internet in remote experimentation. The term *Online Labs* includes both remote and virtual laboratories, provided they are available on the Internet.

Let us look at a remote laboratory illustrating Archimedes' buoyancy principle. We know that some objects float and others do not: a huge metal ship floats and a simple marble sinks. We have physical fact, buoyancy, and to explain it, we will need to throw objects into the water (question) to see whether or not they float (answer). Where these objects come from, how they are thrown and how we observe whether or not they float have no relevance *vis-à-vis* the scientific fact, though they are relevant to the user's experience.

Figure 1.1 shows a remote experiment to test the buoyancy of various objects, first out of and then in the water.

In Fig. 1.1, we can see the controls (arrows) to launch the objects (input variables) while the result of the launch is observed via webcam.

But there is one question still to be answered: why do we do this? What are we looking for? The aim is to understand the present (analysis) or the future (design) in response to three basic questions: What is happening? Why is it happening? And what will happen? Figure 1.2 connects the concepts of analysis and design.

In fact, the previous question depends to a large degree upon the experimenter's goals and on whether he or she is a student or a scientist. One of the goals of basic research is to mathematically model physical and other phenomena; this is the typical work of a scientist. Other people seek to draw on basic knowledge and develop modelling to design technological solutions. This is the typical work of an engineer or developer.

However, the student's goal is usually to test the validity of a model that has already been designed or discover for themselves, via enquiry, the existing model of thought. In other words, the objective is to train and reinforce critical thinking as well as utilise the scientific method.

Scientists and students pursue different objectives, so experiments and laboratories must also differ, whether remote or not.

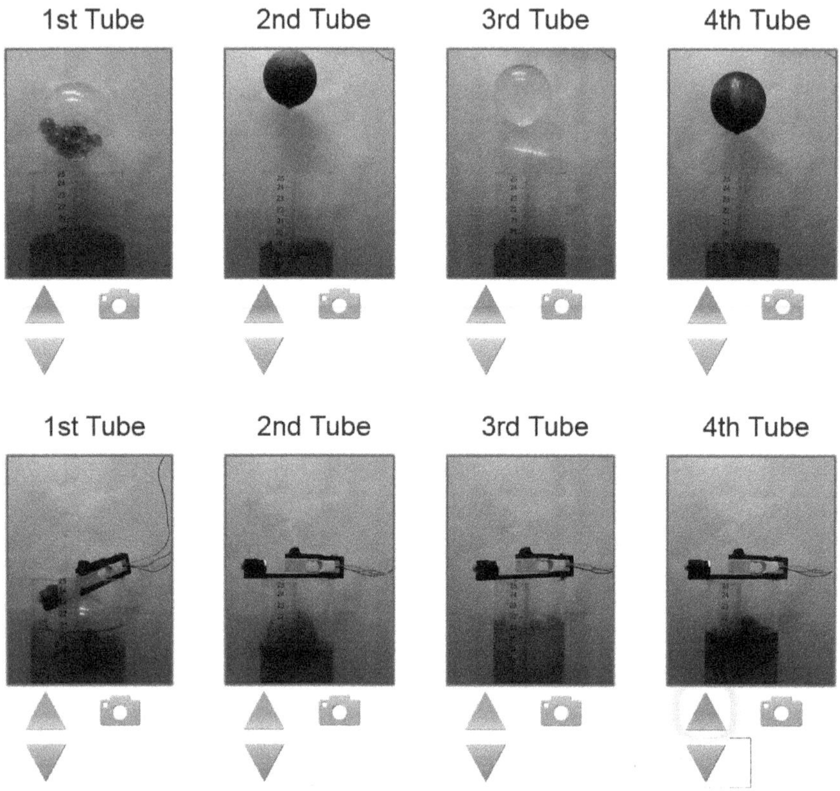

Figure 1.1. Buoyancy remote experiment from LabsLand.
Source: www.labsland.com.

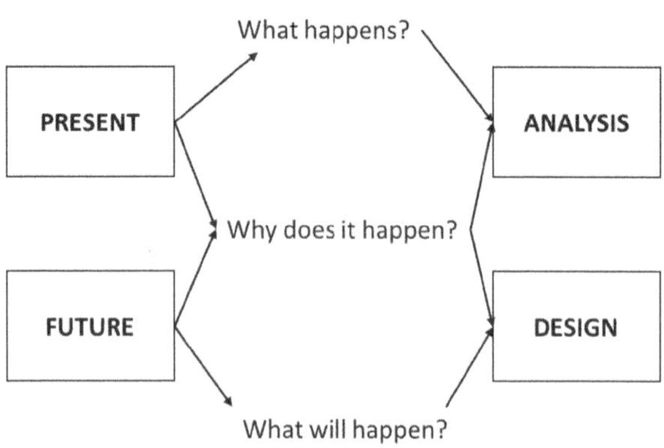

Figure 1.2. Questions, analysis and design.

We need, therefore, to clarify what a laboratory is. A laboratory is not just a collection of measuring and control instruments, but a set of methods and processes with which to obtain useful answers from experiments. Many of these instruments are already computerised and so are, therefore, the associated methods. This situation greatly facilitates the development of remote laboratories since their digital nature is shared with the Internet and its services. Examples of the above are the IoT (Internet of Things), PXI and LXI technologies.

Figure 1.3 shows clearly how a good experiment requires the natural and unforced integration of many elements, criteria and protagonists.

An important question before we continue: Why is it interesting to design and use a remote laboratory? There are many possible answers to this question, and indeed the purpose of this chapter is to tailor that answer to each reader according to their interests and experience. But the first challenge is to address the limited access that students had (and have) to certain experimental educational resources, for reasons of cost, complexity, etc. The most direct and famous answer was provided by Spanish MIT researcher, Jesús del Álamo, 'If you cannot come to the lab, the lab will come to you'. What is clear and should be underlined is the fact that using a remote laboratory in no way precludes (and perhaps encourages) the use of a hands-on or virtual laboratory; each has its own characteristics. Important for J. E. Froyd, editor of *IEEE Trans. on Education*, is what he published in

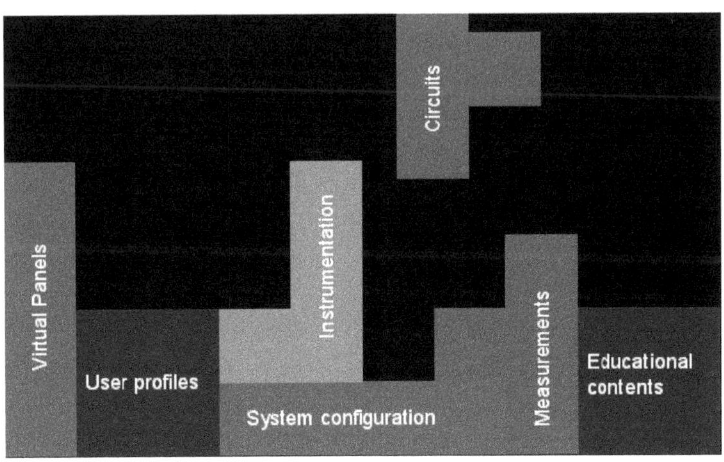

Figure 1.3. Elements of an experiment according to Bagnasco (2007).

2012 with two other authors to commemorate 100 of engineering teaching; the title of the article was clear in this respect: 'Five Major Shifts in 100 Years of Engineering Education'. In the article, the authors describe the five most important axes of engineering education, one of which is the use of TIC in teaching. Within this axis, nine levers are highlighted, one of these being 'Remote laboratories', about which the following is said: 'Although remote labs are not yet common in engineering education, their economic advantages will probably lead to increased use' (Froyd *et al.*, 2012). Remote laboratories form a part of TEL and are a clear exponent of what technology can do for teaching.

Reports concerning education and technology are regularly published. An important question is whether those reports refer to remote laboratories as a crucial educational tool? NMC Horizon Report[1] published an annual report describing the trends in educational technology at different levels, and the 2018 edition included Virtual and Remote Laboratories as a key development for Learning Technologies.

The Centre for Engineering Education is sponsored by 4TU in Holland[2] and in 2017 published the report 'Emerging Technologies in Engineering Education: Do we need them and can we make them work?' (Vries, 2017). This report was based on a meta-analysis of other reports (Gartner, NMC Horizon Report, etc.) in order to choose the four emerging technologies to be addressed: *Virtual and remote laboratories/experiments, Virtual Reality* (VR), *Internet of Things* (IoT), and *Makerspaces and Bringing Your Own Device* (BYOD). The report is very interesting, not only because it highlights remote experimentation as an emerging technology but because it addresses important questions in clear fashion. The report explains that technology is not a good thing by itself, but it is a good thing due to its results in the classroom; the report also highlights the importance of institutions and policymakers. An example of the critical spirit of virtual laboratories and institutions is the recommendation (Vries, 2017, p. 28) 'HEE needs to claim ownership and involvement and not rely on hobbyism or altruism as an excuse for incompetence in the development of educational technology'. In short, the recommendations and the considerations included in this report are a source of inspiration for any person in a position of responsibility in educational technologies.

---

[1] www.nmc.org.

[2] https://www.4tu.nl/en/.

1.2 History of Remote Experimentation

Remote experimentation was started in the mid-1990s, and it was directly connected with Computer-Based Learning (CBL) and with the initial use of the Internet, around 1990. Since the beginning, different researchers in the USA noticed the great impact of the Internet in combination with the computerised instrumentation, that was starting in those days, to create accessible experiments from the Internet: the remote experimentation.

Specifically, the work published by Bohus (Fig. 1.4), Atkan (Fig. 1.4) and Shor of Oregon State University, (USA) in 1996 (Bohus *et al.*, 1996), is the leading work in remote experimentation and was based on a research activity in 1995. In 1996, Jim Henry (Fig. 1.4) of the University of Tennessee at Chattanooga, USA, had published his first work, although he himself explains that he had previously used a remote laboratory in the subject of Engineering Control Systems (Henry, 1996). Jim Henry's remote laboratory is no longer accessible, but its original version contained remote experiments to control processes, in chemistry and mechanics. At the same time and also in the USA, the Cyberlab, designed by Lambertus Hesselink (2003) with Stanford University, is still online,[3] and is devoted to digitised and ultraconcurrent remote experiments.

Even earlier, in 1994, the University of Western Australia, UWA, claimed to have the first remote laboratory in the world. It was designed by Ken Taylor and Barney Dalton (1998), but did not present any publications or work. It focused on robotics and incorporated NI LabVIEW technology.[4] The last update was in 2011.

Worthy of mention in the same sphere is *Telegarden*, designed by Ken Goldberg and Joseph Santarromana, which from 1995 to 2004 made it possible to control a robot gardener. This telerobot was better known in the artistic than in the technological field.[5]

In Europe, Jürgen Kirstein (1997) started a remote laboratory based on filmed experiments in around 1996. This original concept, *ISE (Interactive Screen Experiments)*, is still available (although only in German).[6] However, this pioneering approach is being reactivated by different researchers like Luis de la Torre, (2020).

In the early 21st century, Ingvar Gustavsson (Fig. 1.4), of the Blekinge Institute of Technology, Sweden, developed the *VISIR* (*Virtual Instrument*

---

[3]https://secondilabs.appspot.com/.
[4]https://telerobot.mech.uwa.edu.au/Telerobot/index.html.
[5]https://youtu.be/gbyy5vSg8w8.
[6]https://didaktik.physik.fu-berlin.de/projekte/ibe/.

Systems in Reality) for remote experimentation with electronic circuits (Gustavsson *et al.*, 2009) and is, without a doubt, the most successful functioning remote laboratory during the past 15 years. A little later, the team led by Javier García-Zubía (Fig. 1.4) at Deusto University, Spain, began their activities with a remote experiment that enabled the uploading and control of a Xilinx CPLD within the WebLab-Deusto platform[7] (García-Zubía, 2003). More or less at the same time, in Brazil, Juarez Bento and João Bosco (Fig. 1.4) deployed *RexLab* (Wisintainer, 1999), to coordinate all the remote experiments they had developed since 2000. If the Spanish team was the first to develop a remote laboratory for PLD, the Brazilians were pioneers in microprocessors and, in 1997, developed a remote debugger for the 8051, which was highlighted by Myke Predko in his book *Programming and Customizing 8051 Microcontroller* (Predko, 1999): 'Before ending the book, I wanted to share with you one of the most interesting uses of the Internet I have never seen. The Remote 8051 Debugger allows you ...' Both remote laboratories were desktop applications, true technological heresy at the time.

A major landmark was the start-up of iLAB by MIT, USA, in 1999. Initially, it was headed by Jesús del Álamo (Fig. 1.4), but Judson Harward (Fig. 1.4) was the leader of the research group created in this area (Harward *et al.*, 2004). This laboratory was financed by Microsoft with 25 M$ within the MIT-Microsoft iCampus Alliance until its disappearance around 2013, and during this period it was the largest and best remote experimentation platform, including over 20 federated remote experiments.

At the same time, in 1999 in Norway, a remote experiment similar to the one designed by Jesús del Álamo was designed and described by Shen *et al.* (1999).

In 2006, another world reference in university education, the CoMo research group of Cambridge University (Selmer *et al.*, 2011), set up a remote laboratory (WebLab Cambridge) that covered very different areas, including chemistry, but which unfortunately is no longer operative.

The digital nature of remote laboratories facilitates their incorporation within digital repositories that are made available to users. These repositories usually have semantic labelling and SCORM type categorisation, and in their most advanced version (see Section 2.3.1) form a federation of remote laboratories. Notable in this respect are LiLa (Library of Labs),[8] (Richter *et al.*, 2011) and Lab2Go (Zutin *et al.*, 2010), which are not federations but repositories. The former grouped together a set of remote

[7] https://www.weblab.deusto.es.
[8] https://www.lila-project.org.

Figure 1.4. Pictures of some pioneers in remote experimentation: Carisa Bohus, Burçin Atkan, Jim Henry, João Bosco, Judson Harward, Jesús del Álamo, Ingvar Gustavsson, and Javier García-Zubía (sorry for that).

laboratories that described using the SCORM standard. LiLa represented a list of 9 remote and virtual laboratories, one with the evocative name of 'Remote Farm',[9] at the Technical University of Berlin. LiLa and Lab2Go are not operational at the moment.

Different projects quickly began to be developed and financed. In 1996, the European Union (EU) financed the REMOTE project that included remote control of sophisticated equipment. The EU then financed the DYNACORE project (1998–2000) that applied remote control to telescopes — the D-SPACE project has continued in this field of application. Projects like ReLAX (2000), PEARL (2000), CO-LAB (2001) and eMERGE (2002) were pioneers in Europe in remote laboratories and were not only orientated towards telescopes or other specific equipment. Noteworthy in CO-LAB is the inclusion of a business model in relation to remote experimentation. Other European projects in the sphere of remote experimentation were MARVEL, DERIVE, Lab@FUTURE, REAL, OLAREX, ICo-op, RexNet, EOLES, e-LIVES, VISIR+ and RexNet and PILAR. Worthy of mention too are the platforms LiLa and PROLEARN. The GoLab and NextLab projects currently integrate remote laboratories as a fundamental component in order to promote STEM learning through

[9]https://remote.physik.tu-berlin.de/.

Figure 1.5. Open University OpenSTEM Labs.
Source: http://stem.open.ac.uk/study/openstem-labs.

scientific research. The VISIR+ and PILAR projects develop the potential of the VISIR remote laboratory.

Specifically, University of Deusto has participated in the European projects RexNet, OLAREX, ICo-op, e-PRAGMATIC, NeReLa, GoLab, NextLab, VISIR+, PILAR and e-LIVES via WebLab-Deusto.[10]

One example of operational remote laboratories is the one implemented by the Open University in the UK (Fig. 1.5). Its platform is called OpenSTEM Labs, and, as can be seen in the video, its pedagogical and technical approach is highly complete and complex.[11] Open universities are a paradigmatic example of remote experimentation.

In 1999, USA began the Collaboratories project, (Agarwal and Bair, 2004), directed by Raymond Bair of the Pacific Northwest National Laboratory, USA, with the aim of creating scientific communities that would benefit from sharing technical resources via remote experimentation. Worthy of mention in Australia is the LabShare project (Kostulski and Murray, 2010), in 2008, the first to be backed by a governmental institution; its primary objective was to create a federation of remote experiments, and its starting point was a report on the needs and situation of Australian universities.[12]

The EOLES (Fidalgo *et al.*, 2014) and e-LIVES[13] projects have been financed by the EU and coordinated by the University of Limoges, France, and both focus on the use of remote experimentation in the field of electric and electronic circuits. e-LIVES is a project that combines pedagogy, technology and educational policies and is deployed in countries in North Africa and the Middle East. The report can be downloaded at its website.[14]

[10] https://www.weblab.deusto.es.
[11] https://learn5.open.ac.uk/course/format/sciencelab/section.php?name=golc2.
[12] http://citeseerx.ist.psu.edu/viewdoc/download?doi=10.1.1.470.7265&rep=rep1&type=pdf.
[13] https://e-lives.eu.
[14] https://e-lives.eu/?page_id=10241#.

Nowadays, there is only one company dedicated to educational remote experimentation: LabsLand. This company is a spin-off of the University of Deusto, and its co-founders are researchers from the same university. At the time of writing, LabsLand provides over 30 laboratories (Robotics, Electronics, Physics, Biology, Chemistry) from 24 universities in 14 countries over all continents, and they are partnering with more. LabsLand will be used as a guide in this book, and there will be many examples using laboratories provided by LabsLand.

Recently, in 2018–2019, the University of Hong Kong followed suit with the creation of the spin-off LabWork (https://remotelab.labwork.com.hk/). Currently (August 2020), they offer 15 remote physics experiments, of which only six are available and the rest are "coming soon".

The above-mentioned details might give the impression that there are currently a substantial number of remote experiments available, but this is not the case. The iLAB project was started by MIT in a similar way to LabsLand at the University of Deusto. MIT offered several remote labs in the first decade of 21st century, and then iLAB was granted $25 million by Microsoft to develop more and better remote experiments including the ISA software architecture under the coordination of Judson Harward. However, the iLAB project was stopped in 2013. In Australia, the University of Technology Sydney had a set of remote labs coordinated by David Lowe, and the Australian government granted the LabShare project to create a federation of remote labs. The project was finished at 2013, but at that moment, the company LabShare Initiative was created to offer remote experimentation not only in Australia. Now, LabShare Initiative is not active, but David Lowe is offering remote experiments in the University of Sydney using the Sahara software architecture designed during the LabShare project.

Figure 1.6 includes a selection of current projects.

In the words of Kara *et al.* (2010) 'The majority of these (remote labs) have, however, failed in operating the laboratory beyond its development phase, and making it sustainable'. In the same vein Orduña *et al.* (2016) says 'However, while the number of remote laboratory initiatives is high, the overall impact of these laboratories is fairly limited beyond the scope of the host institution or the scope (and duration) of projects in which the host institution is involved. There are cases where the laboratories are regularly used by other institutions, but these are still exceptions and remote laboratories are not yet widely used. This is not the case for virtual laboratories (simulations), where the maintenance costs and work required once developed tend to be low'. This is a recurring problem in

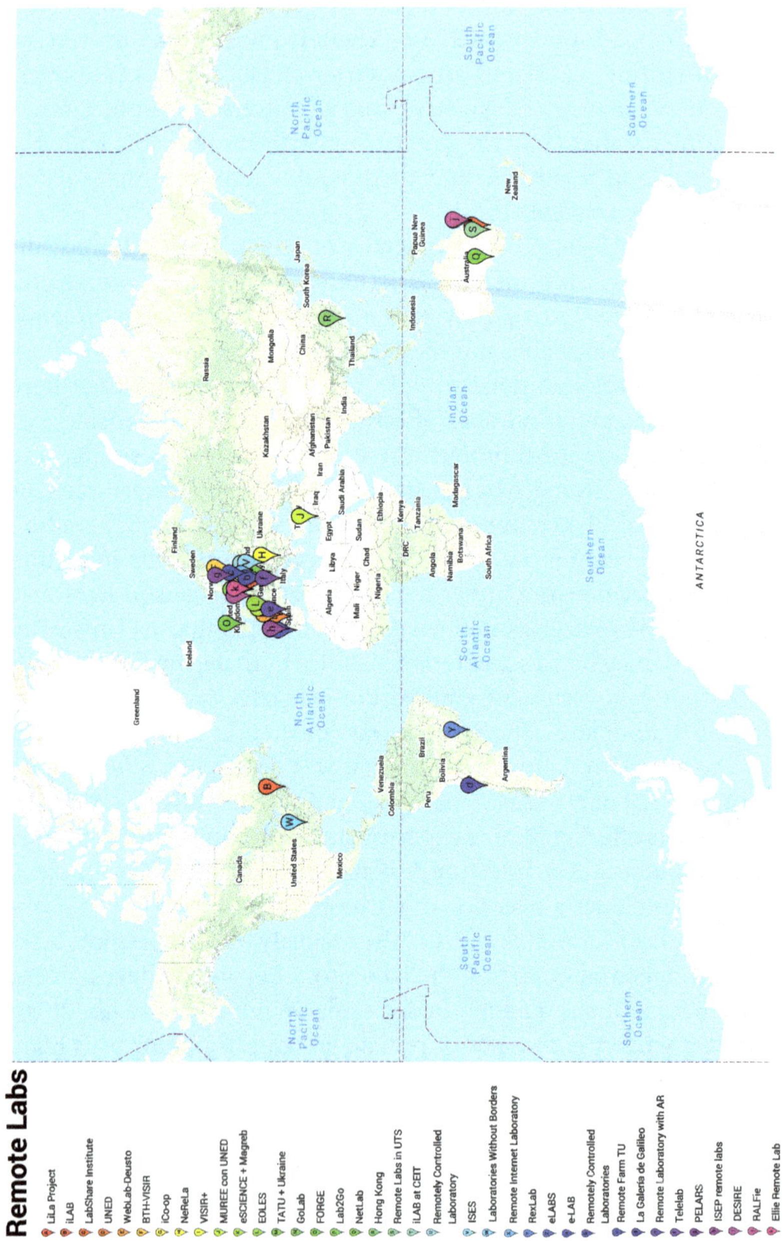

Figure 1.6. Geographical distribution of some remote laboratories.

Source: Google Maps, Map data © 2020 Google, INEGI.

remote experimentation and needs to be addressed from a professional and commercial perspective.

Another interesting aspect from a historical approach is identifying the most relevant research teams. It is not easy to compare teams, as it is not easy to find suitable criteria. A typical approach is analysis of the relevance of each group or researcher's publications; in other words, a bibliometric analysis. There have been two specific studies of remote experimentation: (Zappatore *et al.*, 2015) and (Heradio *et al.*, 2016a).

According to the second study, the most relevant works are those published by Ma *et al.*, Atkan *et al.* and Harward *et al.*, with over 200 mentions of each. Note that three of the researchers with the highest H-index in WoS (Web of Science) are Spanish: Dormido (H-index: 13), Castro (H-index: 8) and García-Zubía (H-index: 7), along with researchers like Gillet, Alves, Gustavsson, Loewe and Nafalski. If the results focus on research teams, the most significant are *UNILabs* (UNED, Spain), *WebLab-Deusto* (U. Deusto, Spain), *iLAB* (MIT, USA), *LabShare* (UTS, Australia) and *UNED* (Spain). Broadly speaking, these results coincide with those obtained by Zappatore *et al.* (2015). The only caveat is that both studies were performed in 2016, and the situation may now be different.

Figures 1.7 and 1.8 show the H-index and the number of papers published by researchers according to Heradio *et al.* (2016a).

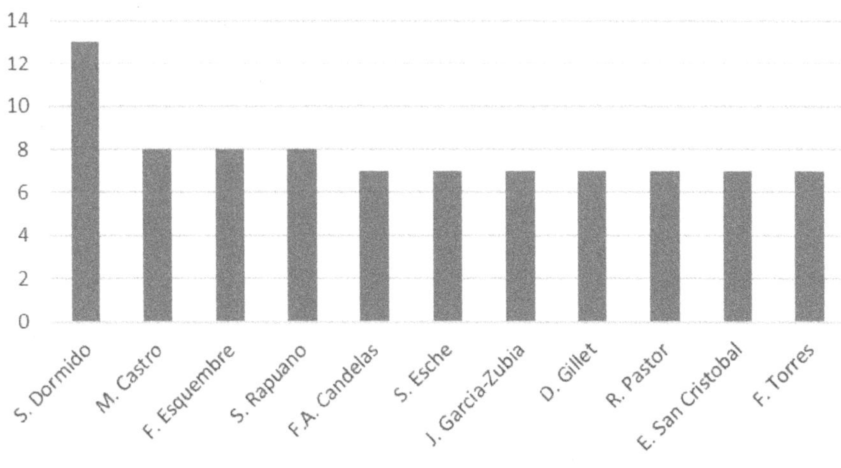

Figure 1.7. WoS H-index for the main researchers in remote labs in 2016 (Heradio *et al.*, 2016a).

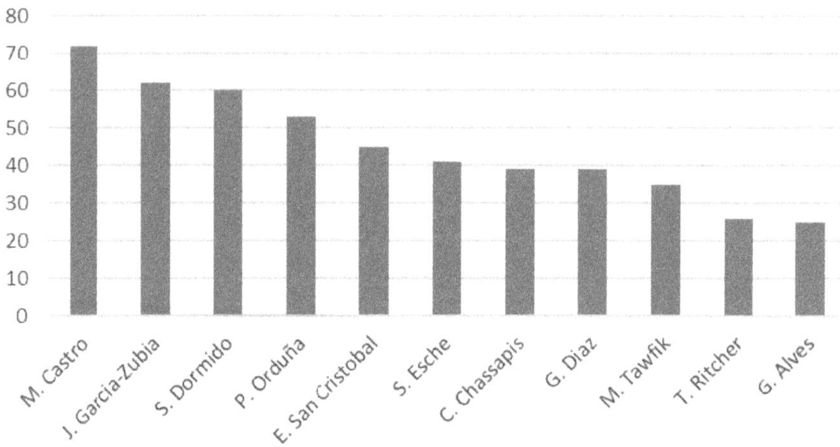

Figure 1.8. Number of published papers per main researcher in remote labs in 2016 (Heradio *et al.*, 2016a).

This detailed analysis has also made it possible to evaluate which journals and conferences are most prolific in the field of remote experimentation. There is a journal called *iJOE,* the main topic of which is remote laboratories. Also prominent within IEEE are *Trans. on Education* (ToE), *Trans. on Industrial Informatics* (TII), *Trans. on Industrial Electronics* (TIE) and *Trans. on Learning Technologies* (TLT). The journals most focused on education and with interest in remote experimentation are *Computers & Education, Computers Applications in Engineering Education* and the *International Journal of Engineering Education.* Regarding conferences, the ASEE Annual Conference, Frontiers in Education, EDUCON, REV and The Experiment@ International Conference stand out, the last two solely dedicated to remote experimentation. In any case, the transversality of remote experimentation makes it possible to publish results in a sizeable number of journals and congresses.

Various books have been published on the subject of remote experimentation. Three of them by the University of Deusto (García-Zubía and Gomes, 2006), (García-Zubía and Alves, 2011) and (Dziabenko and García-Zubía, 2013), the second entitled *Using Remote Labs in Education.* Two of them were edited by Springer (Auer *et al.*, 2018) and (Ko *et al.*, 2004), the first entitled *Cyber-Physical Laboratories in Engineering and Science Education.* Another was edited by IGI (Azad *et al.*, 2012). IEEE-Wiley & Sons edited the book *Lab on the Web*, (Fjeldly and Shur, 2003). Also, the book *Remote Laboratories: in Research-based education of real world phenomena* (Ozvoldova and Schauer, 2016) describes how to teach physics using remote labs. Finally, Restivo *et al.* (2015) relate IoT and remote experimentation. The book *Development of a Remote Laboratory for Engineering Education* has been published in 2020 by CRC Press (Wang *et al.*, 2020). The first book that included a reference to remote labs was *Impact of Advances in Computing and Communications Technologies on Chemical Science and Technology* (NRC, 1999), edited by the National Research Council.

Today, after more than 25 years of work, remote experimentation is a very active field that is advancing in very diverse directions.

1.3 Advantages and Disadvantages of a Remote Lab

We have already defined a remote experiment and a remote laboratory, as well as their role in the classroom for the student. The remote nature of a laboratory — the fact that it is somewhere else and requires an Internet connection — brings both advantages and disadvantages to the educational process, understood as that which involves not only professors and students but also the academic centre.

The list of pros and cons is long, and included here are only those that help to clarify the structure and mission of the remote laboratory. Moreover, some facts that are regarded as advantages by some authors are seen by others as disadvantages.

Below is a list of the most significant advantages for the research community, but as has been noted, this varies considerably from one collective to another and from one person to another — what may be very relevant for somebody is less so for others. The purpose of the list is to provide the teacher and/or the educational centre with criteria with which to analyse the pertinence of incorporating remote experiments as part of their teaching/learning process, without excluding any other form of scientific experimentation.

- **Learning:** Remote experimentation offers students a scientific experience to which otherwise they would not have access, owing to problems related to cost, distance, attendance, etc. The key advantage is having access to what is lacking, albeit not exclusively but in a shared fashion.

- **Availability:** The remote experiment is always accessible, 24/7. A student can always connect to it: in class, at home, at night, at the weekend, alone, with the family, with friends, etc. All that is required is access to the Internet and a computer, smartphone or similar electronic device; there is no need for the teacher to physically open the door of a laboratory.

- **Utilisation:** A teacher may only wish to have one week's access to a remote experiment, and then not use it again, which is not a problem. However, it is difficult to justify an educational centre's investment in a laboratory that is only going to be used for a week. For example, a radioactive experiment is expensive, complex and dangerous, and is very useful in class, but only for a short period of time.

- **Economy:** The educational centre does not need to set up a laboratory (space, staff, equipment, consumables, etc.); it only needs computers and access to the Internet. However, it may have to share maintenance costs and pay some kind of fee for access to the remote laboratories.

- **Collaborative economy:** If an educational centre uses another centre's remote laboratory, it can contribute another to the network, and thus assume costs and share resources.

- **Collaborative didactics:** If several centres or classes are engaged in an identical experiment, then teachers may develop and share common experiences, as well as theoretical content, PowerPoint files and videos.

- **Maintenance:** The teacher or the educational centre does not have to maintain the equipment in a serviceable condition; this is the responsibility of the centre housing the experiment. Only a laboratory teacher knows the work and desperation involved in this task.

- **Student safety:** Whenever students are remote from the experiment, they do not run the risks posed by certain situations that would typically lead the teacher to reject performing these experiments in the classroom or laboratory on the grounds of security.

- **Safety of equipment:** Risk to equipment is reduced as it is computer controlled and not in the students' hands.

- **Time management and student autonomy:** Each student can organise themselves in their own way and at their own pace, which encourages autonomous and meaningful learning — one of the current paradigms in education.

- **Freedom:** Students can repeat the experiment as many times as they wish and do so freely, without being inhibited by the atmosphere in class or the teacher. In a practical class, there are often students (not all) who are afraid to try out and to experiment, which hinders their learning. These students often have the sensation of 'being observed from behind', but in a remote laboratory, the student can work in peace and without fear.

- **Teaching:** If they have a computer in the classroom, teachers can incorporate their experimental demonstration or verification alongside a theoretical explanation without needing to move students to another space and without needing to pause while this takes place. Theory and experiments can be simultaneous.

- **Open innovation:** Remote experimentation is something innovative in the classroom and the educational centre, and the fact that it can 'be taken home' is a way of sharing that innovation not only with the family but with society in general (non-formal education). Everyone can use the same innovative educational tool, without differentiation according to group or level.

- **Enhanced educational experience:** As the experiment is computerised, one can add to it elements of augmented reality or integrate it within a context of virtual reality, and thus create a richer and more complex experiment for the student. For instance, the student will be able to 'see' magnetic fields or the temperature of a liquid, or have an avatar with which to control an experiment.

Similarly, one can list a series of disadvantages.

- **Trustability/availability:** If the remote laboratory 'crashes', or the Internet fails, or the network is of low quality, or there is any other technical problem, then the experiment will fail, and this is bound to result in frustration. This may not be so different from what teachers may come across in their centre, but in the case of remote laboratories the problem and its solution are far apart, and one may not even know where or under whose supervision. This is what may prove frustrating, especially for teachers, who generally prefer not to jeopardise their class or their reputation in front of students. The remote laboratory must be professional, not amateur. An amusing example of this is the message posted in a remote laboratory: *Please note that this facility is offered due to the generosity of staff and students. Therefore, we cannot provide service levels that you might expect from commercial sites. Best wishes.*

- **Queues:** If for an entire class there is only one remote robot, students will have to queue until their turn comes, which may cause frustration. This situation is alleviated if there are various copies of the same robot.

- **Physical safety of the student:** The student does not appreciate the risk involved in performing certain experiments, and therefore does not learn the operational conduct that guarantees safety during the process. This situation has also been described as an advantage.

- **Physical safety of the equipment:** As the equipment is not actually attended, if an unexpected technical problem arises, then the equipment is at risk. For instance, if there is a short-circuit caused by incorrect installation, not only is the equipment at risk but the installation itself or the building too.

- **Suitability for the student:** Is a remote experiment educational? Does it contribute to the learning process? Is it easy to use? Is it comparable to a hands-on experiment? Is it of value to the student? Many of these questions are addressed later in Section 4.1.

- **Suitability for the teacher:** The remote experiment may be too complicated (or simple) for the teacher, and this may require them to run didactic risks when using and showing it to students. An over-designed experiment, a surfeit of information and, in general, an inappropriate level may represent an obstacle to its use in the classroom.

- **Reality:** Remote experimentation always offers a real experience, although the effect caused by distance or the interface may lead the student to observe it as virtual or digital. Nowadays, this digression no longer makes much sense, as young students spontaneously combine the digital and the real, without differentiation or critical discussion.

- **Freedom:** Whenever students access an experiment via a graphic interface, there is a possibility of their controlling it without clear scientific criteria, simply 'pressing all the buttons', as if it were a game or a console. The risk is high because the student eventually achieves a result, 'it runs!!', but not in a systematic fashion. The teacher should assume and minimise this risk.

- **Didactics:** Not all the objectives or the results of learning are attained by a remote laboratory. Some are only attained via a hands-on laboratory. This aspect is discussed in Section 4.1.

Each professor or educational centre involved may assess the advantages in a different way. Some may be relevant and others not, depending on their needs. Something similar occurs with the disadvantages, as one disadvantage may 'outweigh' all the advantages. For instance, if a remote laboratory requires all a school's Internet ports to be opened, then it may be unacceptable to the management, irrespective of the considerable value of the remote experiment or the dimensions of the problem it resolves.

In other cases, like safety, what is an advantage for some is a disadvantage for others. For example, some teachers see that that it is not good for students not to appreciate the risks involved in some experiments since later in real life they will not control those risk. This paradox may arise in other situations.

Though it is not applicable to all the advantages and disadvantages listed, there is a double axis of utility–usability that proves very useful. It is important for an experiment to be useful for teacher and students, but it must also be useable. What is useful or useable may have one meaning from a technical point of view and another from an educational or institutional perspective. Just as in every interesting field of study, the balance between opposites is continuous, and when we win in one aspect we probably lose in another.

It is worth noting that all the advantages and disadvantages listed (and those that the reader could add) have to do with the experiment itself and with the teaching experience in the classroom. They have less to do with the technique employed. But there are technical characteristics that may condition the experience, and these will be considered in Section 2.2.

Another element that arises in this analysis of advantages and disadvantages is that of professionalism. If the teacher decides to complete a remote experiment in class as something innovative but not central, any type of remote experiment may suffice, if one is flexible *vis-à-vis* the usability–usefulness duality. However, if the teacher decides to teach robotics with a remote mobile robot which is going to be their laboratory throughout the academic year, then it is necessary to find a professional supplier (the characteristics are explained in Chapter 5) who not only covers the educational aspects but also guarantees the availability of the resource. Teachers do not want explanations as to why a robot does not work; they want the robot to function in class.

Finally, and after describing the most significant advantages and disadvantages, it is very important to emphasise that a remote laboratory does not compete with other forms of experimentation — hands-on laboratories, virtual laboratories and simulators — but coexists with them, with its own unique disadvantages and advantages. When they are combined, the teacher should seek the maximum impact upon the student in terms of learning achievements, and furthermore should not forget that, on occasion and for different reasons, the remote laboratory is the only means of interacting with nature.

Chapter 2

Characterisation of Remote Experimentation

A remote experiment, RE, from the technological point of view, is conditioned in two ways. First, one has to define the remote experiment itself and, second, do the same with the Remote Lab Management System, RLMS. The RLMS offers the user the remote experiment, and consequently conditions it.

The RLMS and the RE are comparable to the laboratory and the experiment. Even if the experiment designed by the teacher is perfect, if the laboratory is not perfect, or lacks satisfactory working conditions, etc., then the student's experience will be negative or incomplete.

The same thing occurs with LMS (Learning Management Systems) like Moodle, Sakai, Google Classroom, etc. and the materials and activities to which they provide access. If Moodle is not operational for a teacher, then neither are the activities it contains, however perfect they may be.

Therefore, we will first describe remote experiments and then RLMS, so the reader is clear as to what each involves.

Then we will classify different examples of RE and RLMS according to the criteria explained below.

2.1 Types of Remote Experiments

First we have to compare 'experiment', 'remote experiment' and 'remote laboratory'. The latter may offer the student more than one remote experiment, or just one, in which case the experiment and the laboratory merge. This section describes experiments per se but seen as if they were going to be remotised or already were. Section 2.3 specifically studies laboratories and remote experiments.

There are multiple criteria and points of view regarding which experiments can be classified. One criterion is by scientific area: physics, chemistry, biology, programming, etc. Another is differentiating between remote experiments that are easy or difficult to use, design, etc. For example, carrying out a biology experiment is very complicated, since one has to succeed in manipulating living organisms via the Internet, whilst experimenting with an electric circuit may be very easy. This is why there are many remote experiments involving electricity or electronics, and few involving chemistry or biology. The more manual intervention an experiment requires, the more difficult it will be to remotise, as 'those hands' will have to exist and be controlled remotely. Think of liquids that have to be poured into test tubes and then mixed together: how is this remote handling achieved? By using a robot? Are the economic costs and the technological efforts justifiable for each specific experiment? The remote experiment designer has to answer this question.

Classifying a set of elements necessarily involves in-depth analysis thereof, since in order to classify them first we have to find the most relevant criteria and then apply them. Classification is a relevant result, but more relevant still is the actual classification process. In other words, when reading this section, the classification itself is not as important as the elements presented during the classification. The description will enable the reader to begin to understand both the potential and difficulty implicit in remote experimentation.

In this section we will describe and classify remote experiments from different points of view, but the key element of this section will be the classification of remote experiments according to their nature, so technological aspects are excluded from this classification. In this first classification, the noun 'experiment' is more important than the adjective 'remote'.

		NATURE OF THE RESOURCE	
		Real	Simulated
ACCESS TO THE RESOURCE	Local	Hands-on lab	Mono-user virtual lab
	Remote	Remote lab	Multi-user virtual lab

Figure 2.1. Classification of remote laboratories according to Dormido (2004).

2.1.1 *Nature of the experiment*

Taking the discussion into a more academic field, first we can consider the classification performed by (Dormido, 2004) and shown in Fig. 2.1 in order to situate the remote laboratory. Initially, a remote laboratory offers the user the possibility of completing a real experiment without actually being present.

In the top left, we can see the hands-on laboratory where the resource is real and in front of the student. Below, the remote laboratory differs from the former only insofar as the student is not sitting opposite the resource, but with the Internet in between.

A simulator reproduces reality by using a logical-mathematical model programmed in a computer. In a simulator, the choice of interface is relevant, since if this is sophisticated (virtual or augmented reality) the student may have the sensation of being immersed in reality, even if none of the data is real. This book does not address simulation.

Rodríguez-Gil *et al.* (2017) adapt Dormido's (2004) initial approach to propose the classification in Fig. 2.2. A hybrid experiment is partially real and partially simulated.

An example of a hybrid laboratory is the one developed by (Rodríguez-Gil *et al.*, 2017). This experiment gives the student control using an FPGA (a type of electronic controller) programmed in VHDL (Very High Speed Integrated Circuit Hardware Description Language), a water tank with one valve, two pumps and several level sensors. The FPGA is real, and in it the students load a programme in VHDL written by themselves. This programme accesses the water tank sensors and by means of the latter controls the state of the pumps, which modifies the level sensors and therefore may provoke a new action on the part of the FPGA (Field Programmable Gate Array). In this remote experiment, the deposit is simulated, not real: there is no water tank with thousands of litres of water and a pair of expensive pumps consuming energy.

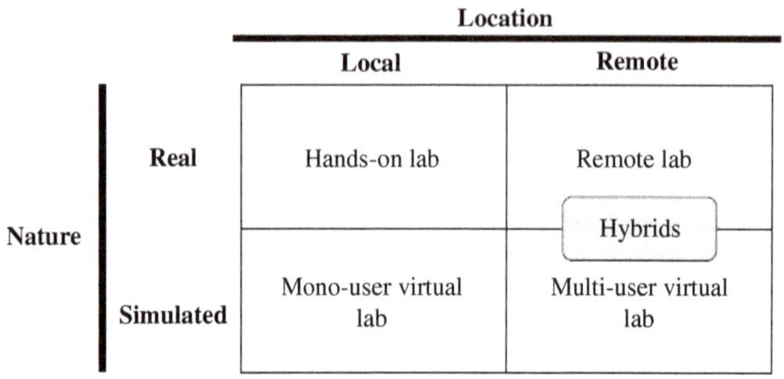

Figure 2.2. Extension of the classification of remote laboratories according to Rodríguez-Gil *et al.* (2017).

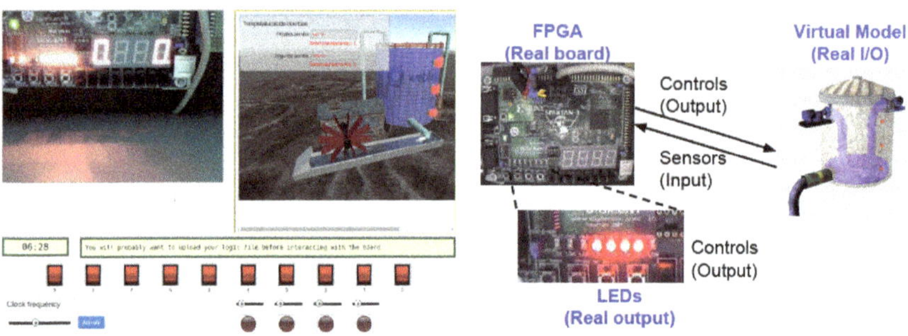

Figure 2.3. Example of a hybrid remote experiment with FPGA and VHDL (Rodríguez-Gil *et al.*, 2017).

This solution is not real, but it costs far less to create and maintain and, moreover, involves no operational risks. In this case, the simulation is a simple piece of equipment, but it could represent a nuclear reactor or an industrial plant.

This kind of remote experiment, see Fig. 2.3, stands out when the most important factor is the student's task of programming the FPGA (or designing the tank, nuclear plant, etc., controller), more than the process to be controlled.

The use of these technologies is addressed later in Section 7.2.

2.1.2 *User interaction with the remote experiment*

According to Harward *et al.* (2008) of MIT, there are three ways in which the student can interact with the remote experiment: batch, interactive or online and sensorised. This depends on the manner in which data is obtained and analysed or, in other words, on the form of interaction with the experiment.

These criteria can be analysed from two points of view:

- How the experiment data is obtained or how it is interacted with:
  - **Interactive or online:** Students continuously interact with the experiment via the interface, change its inputs and see the results in real time. The interface and the webcam are relevant.
  - **Batch:** Students prepare the experiment on the interface and 'launch' it. After a short time (seconds or more), they receive the results. The webcam is not usually important as there is no visual information.
  - **Sensorised:** Students simply monitor and analyse data provided by sensors but under no circumstances influence the execution of the experiment. It is a subtype of the online version, but without interaction.
- How the experiment data is analysed:
  - **Online or in real time:** At the precise moment when data is generated, the student analyses it.
  - **Offline or in deferred time:** The results of the experiment are analysed after they are produced, maybe hours, days or years later. This kind of experiment is known as Data Set. These experiments are also called pre-recorded or ultraconcurrent.

If the experiment designed by the teacher is interactive, then it will be online. For example, in the case of buoyancy, the student has access to various objects to launch and must decide in real time which objects to launch and what to expect from these: whether or not they float. The learning experience designed by the teacher requires the student to interact with the remote experiment.

However, the teacher may have addressed buoyancy from another angle: the student gives the order to launch all the objects, and the remote experiment responds with an image (or a file) indicating which objects have or have not floated. In this case, it is a batch experiment: once launched, interaction with the student is not necessary.

In another example, one might wish to characterise the behaviour of a diode (or its resistance) by measuring the current passing through it when voltage is applied. This process is then repeated for a considerable number of various voltages.

This process can be executed one-to-one (online), or one can create a set of voltages which are then sent to the experiment. The laboratory powers the diode with the entire set, records the currents one by one and returns them all together in a file or an image. In other words, this is a typical batch process. In this example, the second mode appears to be better, but it is the teacher or designer who decides to use one or the other.

The UNR remote laboratory (Lerro *et al.*, 2008) (Fig. 2.4) obtains the characteristic curve of a diode as a batch process, while VISIR does so in online mode, point by point.

Let us consider the case of robot controlled with an Arduino (https://www.arduino.cc/) programmed by the student. If, for example, the algorithm created by the student allows them to press an arrow on the interface and the robot must move in that direction, then the remote robot will be online, as the student has to interact with the algorithm to see whether or not it is correct.

However, if the goal of the algorithm is to make the robot follow a line (see Fig. 2.5), then the robot may be online or batch. Normally, the student wants to see in real time, online, how their robot behaves in order to see whether or not the algorithm is correct. But the possibility also exists that the robot does not provide this option and, in this case, the sequence is as follows: the student improves the algorithm now and, later, perhaps overnight, the system uploads the algorithm in the robot's Arduino and executes it, making a video recording of the robot's behaviour. The following morning the student receives the video and sees whether or not the robot has followed the line and decides whether or not to change the algorithm.

Clearly, the online option seems preferable, but the batch option may be more efficient. This is a decision for the designer that will affect how students work and most probably their learning, so the teacher may or may not agree with the online or batch mode chosen.

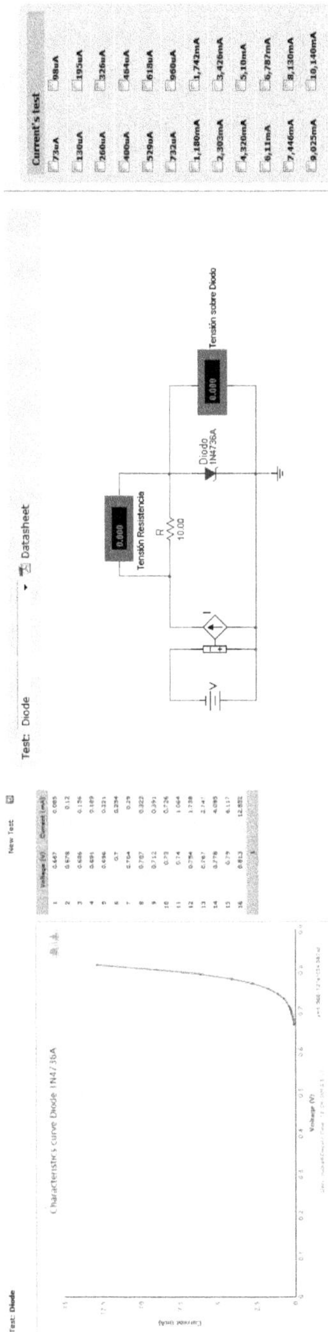

Figure 2.4. Example of remote batch experiment.

Source: https://labremf4a.fceia.unr.edu.ar/.

Figure 2.5. Example of a real-time online remote experiment with a programmable robot via Arduino.

Source: www.labsland.com.

On the basis of the previous examples with diodes and robots, there are two types of interaction:

- If students interact during the experiment and their interactions modify the evolution of the latter (for instance, heating the water in the Archimedes' experiment), then it is a case of interaction.
- If students can only parameterise and launch the experiment, then interaction is low, as it does not condition the experiment during its evolution.

All interactive experiments have to be online, so students can see in real time the effects of their interaction, but an experiment may also be online without a high degree of student interaction.

2.1.3 *Observability and controllability of the remote experiment*

Another classification is that which considers the type of inputs and outputs available during remote experimentation. The input and output variables can be observable and controllable; in other words, one can access them and know their value (observability) as well as change their value via direct or indirect methods until they reach a known state (controllability).

- Input variables should be controllable by the student, and thus they can choose which values to take at any given moment.

Of course, input variables should be observable, i.e., the student should know their value via the interface of the remote experiment.

- The output variables should be observable; in other words, the student should be able to see on the interface the values of these variables.
- The output variables will be controllable or not depending on the type of experiment.
- In the scientific method, one of the input variables is the independent variable, while the rest of the input variables are fixed, and the output variables are dependent variables.

This approach coincides with Dieter Müeller's classification, which referred to read/write depending on the possibilities of interaction and control offered by the remote experiment. Dieter links both characteristics, read/write, to the security and complexity of the experiment itself, as is shown in Fig. 2.6.

Of course, not all input variables are controllable (write), nor are all output variables observable (read), neither in a remote nor in a normal laboratory.

For example, the buoyancy experiment can be described in Table 2.1.

Remote Sensing ('read'): Acquiring information about system behaviour: · Remote observation · Remote monitoring · Remote measurement · Remote recording ·	MINIMAL RISK AND SAFETY PROBLEMS
Remote Action ('write'): Changing of system behaviour from remote: · Remote parameter assignment · Remote programming · Remote manipulation of physical objects ·	INCREASING RISK AND SAFETY PROBLEMS

Figure 2.6. Classification of remote experiments according to Müeller and Erbe (2007).

Table 2.1. Analysis of the buoyancy remote experiment from LabsLand.

Name and description of variable	Type of variable	Controllability in the LR	Observability in the LR
Objects	Input	Partially controllable: 16 objects	Observable via webcam
Shape of the object: spherical, cubic, pyramidal, shapeless, etc.	Input	Only spherical Non-controllable	Observable via webcam
Mass and weight of object out of water	Input	Partially controllable: 5 masses/weights	Observable via interface
Volume of object	Input	Partially controllable: 5 volumes	Observable via interface
Density of object	Input	Partially controllable: 5 densities	Observable via interface
Liquid receptacle	Input	Test tube only Non-controllable	Observable via interface
Type of liquid	Input	Partially controllable: 3 liquids	Observable via webcam
Volume of liquid	Output	Non-controllable	Observable via interface and via webcam
Position of object in the water, % of volume inside and out	Output	Non-controllable	Observable via webcam
Weight of object in the water	Output	Non-controllable	Observable via interface
Pushing force of the liquid	Output	Non-controllable	Non-observable

As can be seen from Table 2.1, as simple an experiment as the buoyancy one requires control and observation of many variables, but many of these are not relevant to the experiment, such as, for example, the shape of the object or of the water receptacle. It should be borne in mind that the classification of variables as important or not is part of the design of the experiment. This experiment, according to Müeller and Erbe's (2007) classification, is a clear example of 'read'. Other remote laboratories can be analysed using the same approach.

A good experiment should be rich in variables, but should also be simple and usable. This balance is what makes an experiment interesting,

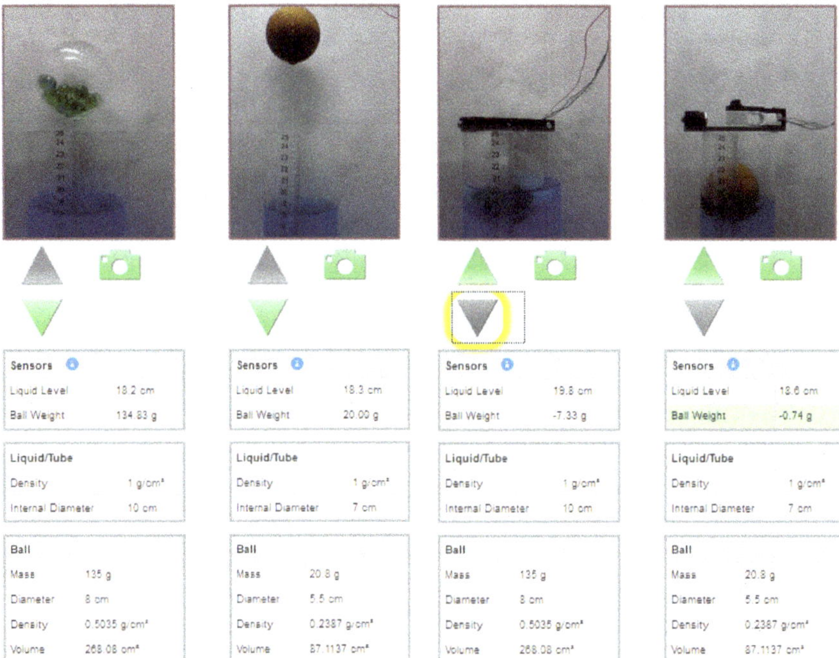

Figure 2.7. Buoyancy remote experiment.

Source: www.labsland.com.

remote or otherwise. If the experiment is remote, then this design is the responsibility of the design team, but if it is hands-on, this design is the teacher's responsibility.

Figure 2.7 shows two laboratory objects before and after being launched into a liquid (water in this case). In it, one can identify some of the different variables from the previous table.

Returning to the input and output variables, the latter should be observable as otherwise the experiment would be incomplete or incorrect. The controllability of the different input variables should be decided during the design of the remote experiment. According to García-Zubía *et al.* (2006), the remote experiment in its design may be characterised by the following:

- **Non-controllable:** The student can only press the ON button, can only initiate the experiment and cannot modify the value of any input variable. The student sees how the experiment evolves and how the output signals change. This is similar to the sensorised experiment in Harward's classification.

- **Parametrizable:** The student can choose for an input variable a value from a set of values. They may parametrise the experiment, but cannot change its structure. For instance, choose the value of a resistor, modify the speed of the motor of a mobile robot or choose from a selection the value of the constants of a PID regulator. Once this has been done, or each time the parameters are changed, the experiment is like the previous one.

- **Logical or structural:** The student can change the structure or the logic of the experiment or perform various different experiments with just one remote experiment. For example, assemble a different electric circuit each time (VISIR), freely programme a mobile robot or an Arduino or design the algorithm of a PID (Proportional-Integral-Derivative Controller). Each experiment may be totally different from the previous one.

In this classification, in the first case, the student can only see what is happening and cannot modify the phenomenon being observed. In the last case, however, the student can create or modify the experiment in such a way that the behaviour of the output variables will be controlled (or not) by the student.

The first type is the simplest of all ('read'), while the last is the more complete and complex ('write'). Let us not forget that a remote experiment is not better on account of being more or less complex. It is the particular nature of the experiment that should establish the complexity of the design. In any case, the designer will certainly attempt to fulfil the objective with maximum possible simplicity.

2.1.4 *Control of the phenomenon of the remote experiment*

In this case, what is important is the user's position *vis-à-vis* the experiment: whether they observe or control.

For example, in the buoyancy remote experiment, the student launches objects and sees whether or not they float. The same thing could happen with an electric circuit in which the user changes the supply voltage to measure the current and thus checks Ohm's Law. In these cases, the student organises the experiment, observes it, and records and finally analyses data. The student's fundamental task is to understand a phenomenon of nature via a scientific approach. The mental work is *a posteriori*.

In the second type of remote experiment, the student does not have to understand a phenomenon — they have to control it. For instance, if the student has understood how a PWM (Pulse Width Modulation) signal controls a DC motor, then they can programme an Arduino to generate the appropriate PWM so that the motor runs correctly. The student's task is *a priori*: first, think how to control the motor and then write the algorithm. They then test the validity or invalidity of the algorithm in the remote experiment. The student may have to initiate this experiment several times until the algorithm functions correctly. However, experiments of the previous type are only repeated if the student wishes to observe the experiment's behaviour again to record a value that may have been forgotten.

Experiments of the first type are generally scientific, while those of the second type are mainly technological. The former seeks to understand a phenomenon, while the latter seeks to control that phenomenon, after understanding it.

This classification is directly related to the previous one. A read experiment is not controlled while a controlled experiment must be of the write type, whether via its parameters or via its logic.

2.1.5 *Combinational or sequential experiment*

If in a remote experiment the behaviour of the output at t only depends on the behaviour of the inputs at t, then the experiment is combinational: the different combination of inputs generates the possible states of the outputs. Their behaviour can be predicted *a priori*, above all if the number of input combinations is not very high.

If in a remote experiment the behaviour of the output at t depends not only on the input in that instant but also on the output at the previous instant, $t-1$, then the experiment is sequential and passes through different states. In this case, it is very difficult to predict the behaviour of the system unless the experiment passes through a few states. A sequential system behaves, in the best-case scenario, like a deterministic finite automaton.

Combinational system	$y(t) = f(x(t))$ *where x is the input, y is the output*
Sequential system	$y(t) = f(x(t), y(t-1))$, *where x is the input, y is the output*

An observable, non-controllable experiment, is combinational in general. In fact, a controllable system can be combinational, but the control strategy is so that it is unlikely to give rise to a didactically interesting controllable remote experiment.

A controllable is usually a sequential system as the control is based on knowing the current state of the experiment, $y(t - 1)$, to 'take it' to the desired $y(t)$ state. The controller can be an algorithm software, or a PID hardware regulator, or an electronic circuit whose set of states will be defined by the student, and therefore it is difficult to estimate the number of states or the relationship between the latter. Thus, although the automaton is finite and deterministic in its evolution, it is impossible or very difficult to predict its behaviour *a priori*.

According to the read/write classification, a sequential experiment will be 'write', while a 'read' type experiment will normally be combinational. The sub-type 'parametrizable' occupies the diffuse zone of those experiments that are 'write' and combinational, and that of those which are 'read' and sequential.

2.1.6 *Summary*

To summarise, a remote experiment is classified as follows:

- The nature of the phenomenon can be physical, chemical, biological, electronic, etc.
- The phenomenon can be real or simulated.
- The experiment and its inputs and outputs can be observable (read) and/or controllable (write) via parameters or logic.
- The experiment may or may not have a controller.

- From the designer's perspective: It can be batch or online.
- From the user's perspective: It can be passive, interactive and batch.

All these criteria mean that the evolution of remote experimentation is very diverse and heterogeneous. An interesting activity would be to synthesise and relate different types of remote experiment since they are not mutually exclusive, but do condition one another.

Everything in this section refers to the remote experiment per se, not to its implementation. However, some features condition the implementation of the experiment, and the latter condition its usefulness in the classroom. Therefore, the students' final experience will depend on the students themselves, the teacher and the designer. These aspects are addressed in Sections 2.3 and 2.5.

It is also important to emphasise that, in general, one type of experiment, remote or hands-on, is not better than another, but may be more or less suitable depending on the kind of experience and the student. The teacher and the student are the central elements in this discourse since a good remote experiment is one that produces learning results in line with the curricular objectives of the subject or degree, provided the designer does not condition them.

2.2 Parts of a Remote Lab

This section describes the different parts of a remote laboratory, from both a descriptive and a technological perspective. Emphasis will be placed on the former criterion, as the latter will be considered later in Chapters 5 and 6.

2.2.1 *Parts and operation of a remote laboratory*

A remote experiment enables a user to control a technological or scientific experiment via the Internet.

There are two phases to the user's experience:

1. Access the remote experiment: Remote Lab Management System.
2. Control and observe the remote experiment: Remote Experiment.

Considering the first phase is not obligatory but is recommended, the student must introduce some form of accreditation (user/pass) in order to access a rare resource. Sometimes access is so simple that both phases merge.

To the aforementioned confusion from the user's point of view is added the chaos caused by some designers that have followed the following steps:

1. First, they set up an electronically controlled experiment.
2. Then they remotise it.
3. Finally they add the access web page.

This bottom-up approach is typical of an amateur designer of remote experimentation, in which hardware skills generally prevail over software design. However, if the goal is for remote experimentation to be present in the 21st century classroom, then the service should be professional. Experiments and their control are important (hardware), but so is the way in which they are accessed (software).

As yet another example, Moodle makes it possible to upload a variety of educational material, share it, re-use it, etc. The focus is not on the material but on transferring it to students and facilitating teachers' work. This is what a professional remote laboratory should aspire to.

An RLMS is the Moodle of remote experiments. If there is only one experiment, then RLMS is not necessary, but if there are several experiments, then it is very important. Access is managed by the RLMS, not each RE; student queues are managed by the RLMS, not the RE; having several similar REs is managed by the RLMS; sharing an RE between various schools is managed by the RLMS, etc. The idea is very simple: everything that is common to any experiment offered must be implemented within the RLMS.

An RE is only responsible for the experiment itself and its remotisation, isolating itself from the rest. The fundamental parts of a RE are as follows:

- **Sensors:** Measure the experiment's input and/or output signals, usually electronic.
- **Actuators:** Control the experiment's input and/or output signals, usually electronic.

- **Controller:** Is responsible for controlling the behaviour of the experiment via an algorithm controlled or not by the user. It does not always exist.

- **Experiment:** The pendulum, the robot, the object in the liquid, the electronic circuit, etc. The object of the experiment.

- **Webcam:** Offers the user an image of the behaviour of the experiment. It may be seen as a specific sensor that does not always exist.

- **Web interface/User interface:** Enables the user to see the values of the sensors and the webcam video, control the actuators and change the algorithm of the controller, if the latter exists.

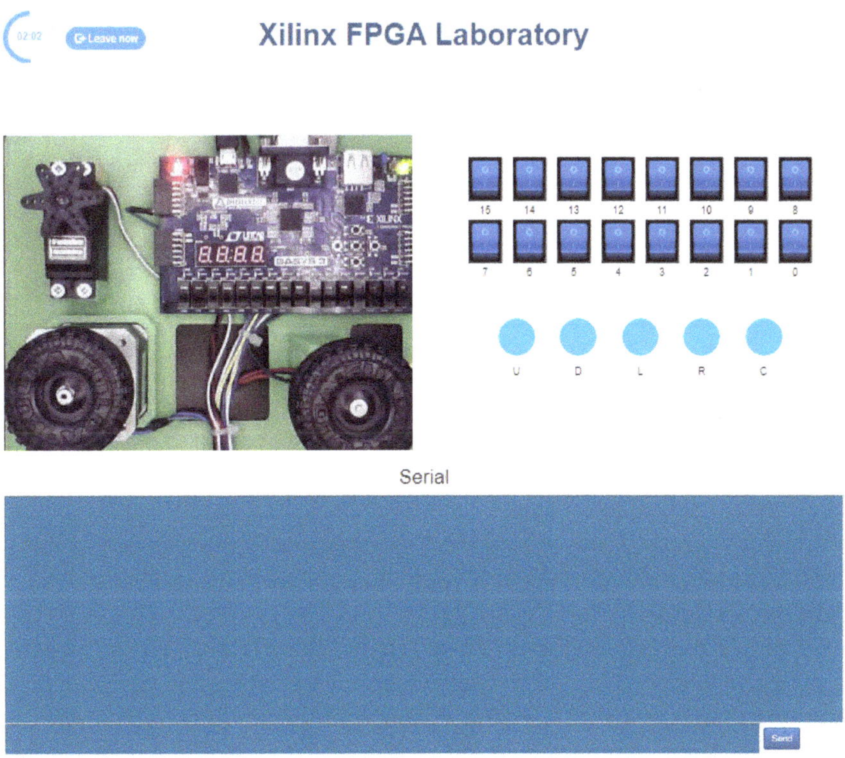

Figure 2.8. Remote experiment with Arduino.

Source: www.labsland.com.

It is not necessary for an RE to have all the previous elements, but they are very common. In many cases, the controller is not necessary, as we will not want to control a phenomenon but simply see it. We have already mentioned the fact that in remote laboratories with scientific experiments there are normally no controllers, while in most technological remote laboratories (robots, Arduino, FPGA, ...) the controller is the main element as the student's activity consists of just creating an algorithm. See examples in Section 3.2.

Figure 2.8 shows the interface of an Arduino remote laboratory where the student can create algorithms that control motors by means of buttons and switches.

2.3 Features and Requirements of Remote Laboratories

As seen in the previous section, it is necessary to distinguish between remote experiment and remote laboratory. In the former, what predominates is the control of the experiment via hardware–software equipment, while in the latter what predominates is the software platform that offers these remote experiments to users. This platform is known as Remote Lab Management System, RLMS.

The way in which the experiment is accessed depends on the RLMS, while the experience depends directly upon the remote experiment. Historically, more importance has been given to the experiment than to the RLMS, but now the RLMS totally conditions the user's experience.

Remote laboratories will have attained an initial degree of acceptance when they are offered by schools, universities, publishers, etc., as this would mean they are necessary and reliable, being of equal importance. No institution is going to incorporate into its everyday operations something that is not completely reliable.

In this section, we will address the requirements (satisfied or not) applicable to an RLMS and to a remote experiment in order to render them professional.

Although it is sometimes not easy to differentiate between the remote experiment and the laboratory, requirements are addressed in the following sections from the twofold perspective of the RLMS (how experiments are accessed and maintained) and the RE (how it is implemented). These requirements guarantee accessibility and sustainability.

2.3.1 *Requirements of an RLMS*

If we said earlier that advantages and disadvantages referred to the remote experiment, to classroom activity (*a posteriori*), the requirements in this section refer to the design of the remote laboratory (*a priori*), to the way in which those remote experiments are 'taken' to the class, the students and the teacher.

The characteristics/requirements are described in detail later, but before that it makes sense to group them into the following four concepts:

(1) **Universality and accessibility:** Describes whether a laboratory is accessible in any technological scenario in the classroom or educational centre, or even outside it.

(2) **User management:** Describes how users access the remote laboratory and the record of their activity.

(3) **Scalability and extensibility:** Describes how easy it is to extend the remote laboratory by adding new experiments or replicating existing ones.

(4) **Learning support:** Describes whether the remote laboratory facilitates the work of the student and teacher in class.

Herein, we provide a detailed description of the four previous categories that should be seen as requirements in the design of a remote laboratory attributable to the RLMS (later we will refer to how they are achieved in Chapter 5):

(1) **Universality and accessibility**

- **Trustability:** The laboratory is available and is not 'out of service', 'under maintenance', etc. Unfortunately, it is very common for a remote laboratory, after its design, to cease operations in a matter of months.
- **Universality:** A remote laboratory should be accessible from any kind of computer, tablet, smartphone, etc. It should work

with any operating system and any browser. This feature is self-explanatory as students do not want to find themselves overly dependent on different technologies.

- **Installation:** A remote laboratory should be accessible without the student or teacher having to install any additional software, any plugin or any other requirement that might condition their level of safety or the policy of their educational centre (in many centres, the installation of external programmes is specifically prohibited).

- **Security:** Access to a remote laboratory should under no circumstances compromise security in the educational centre by requiring certain insecure ports to be opened, certain firewalls to be deactivated, etc. Security must always come before utility.

- **Efficiency**: It is vital for the remote laboratory to consume the minimum possible resources without detriment to the quality of the experiment in question. Bandwidth is the most obvious example: the less that is consumed, the better, since in many educational centres there are restrictions in this respect. Let us remember that the video is a fundamental element of the remote experiment; it is the way in which students see how their experiments evolve.

(2) **User management**

- **Management of usernames and passwords:** If access to a remote laboratory requires a username and password, the best thing is for students to be able to do so using their educational centre passwords by means of a secure method. It is not good practice to ask students to register at a new domain and use different usernames/passwords, and this is certainly undesirable when students are below a certain age.

- **Queue management:** In an ideal case, whenever a student wants to access a remote laboratory they should not have to wait, there would always be an experiment free, but in the real case if there is only one experiment for all the students in all the centres, then they will probably have to wait in a queue. Appropriate and flexible queue management by the remote laboratory is one of its main characteristics, and of course, availability of various examples of the experiment itself. One alternative to queues is to use a booking system. In this case, the student books a time slot, so nobody else can access the experiment, which guarantees its availability for that student, but reduces availability and efficiency for students as a whole.

- **Data collection and learning analytics:** It is increasingly important for teachers and the educational centre to be able to observe and know what use students make of the remote experiment: how many times they connect, when, for how long, what they do, etc. The students themselves appreciate the fact that the teacher can see their evolution. The data obtained will be analysed by the teacher.

(3) **Scalability and extensibility**

- **Extensibility:** A good remote laboratory should not be designed for one single experiment, but for several. The ease with which a new remote experiment is integrated is what is described by the extensibility of the laboratory. The plug&play paradigm is not currently available on any platform.

- **Scalability:** It is not easy for the student or the teacher to understand scalability, but it is crucial as it determines the ease with which one can add copies of the same experiment, in other words, it refers to the facility with which it serves more students and educational centres.

- **Federation:** If an experiment is replicated in different places, for example in Spain and England, it may be the case that the students in Spain only use theirs, and those in England only theirs, or they may be federated. In this case, each group of students will have access to two experiments, not just one, which increases availability and eases the queue. Technically, this is a complex feature to incorporate and involves other features such as load balancing, error response, use tracking, etc., without forgetting the user agreement that partners must sign.

(4) **Learning support**

- **Learning platforms:** Teachers are making increasing use of LMS like Moodle, Google Classroom, etc. as platforms to improve students' learning. It is fundamental for a remote laboratory to allow for the integration of its experiments in Moodle, for instance, so teachers can organise all their work material in one single platform (the educational centre's), without obliging the student to enter others.

- **Didactic support:** The teacher should be able to consult and debate the didactic usefulness of the accessed remote experiment. For example, they may ask for clarification with regard to its use or even have doubts *vis-à-vis* the actual fact to be tested. A teacher will hardly ever demonstrate in class something about which he is not certain; only exceptionally do teachers take risks in front of students.

- **Technical support and maintenance:** What happens if a robot gets stuck at some time because it has been incorrectly programmed? What happens if a pendulum breaks when it is launched? Provided an experiment is real, the previous situations can arise, and in this case the important thing is that the laboratory supplier has response capacity: has several copies of the experiment, provides a technician 24×7, answers technical queries, etc.

- **Price and sustainability:** The best way of ensuring a quality service is by paying for it. This also ensures its sustainability. It is normal for universities to offer their remote laboratories free-of-charge, but they are not usually sustainable in the long run. This is clearly not a technical requirement, far from it, but it conditions the whole experience.

We will now explore some of the above points by considering examples of remote laboratories.

If a remote laboratory is not reliable or is usually 'down', then the laboratory is not trustable. Thus we move from a technical requirement to a negative frame of mind. Reliability is a prerequisite to the use of a remote laboratory. Is the teacher going to present students with an experiment he cannot trust because he does not know whether it is available? Is a teacher going to want to use a remote experiment which depends on factors beyond her control? Is it acceptable for an experiment to be accessible in one place (computer-network) and not in others? For it to work at home but not at school? It seems obvious that the answers are no. Trust is difficult to win and easy to lose, and teachers are very demanding in this sense.

For instance, one of the best remote laboratories is the OE@FEUP in Porto. Figure 2.9 warns of a possible problem with Internet Explorer.

The prestigious Open University, before granting user access to some of its remote experiments, checks whether the user is going to experience any access or connection problems so as to avoid them suffering a negative experience. Figure 2.10 shows a refusal of access to an Open University remote laboratory from the University of Deusto network.

The question now is knowing where the problem is. Is the University of Deusto network incorrectly deployed? Should it open more ports or remove firewalls? Is the remote experiment badly designed, as it requires those ports to be opened or firewalls to be removed? Who should give way? Is this debate of interest to the teacher or student?

In a similar vein, the company LabsLand offers remote laboratories set up in universities and research centres unrelated to the company, and therefore it is crucial that it has technology capable of detecting whether a remote experiment is trustable and to what extent, in other words, if it is offline at any given moment, how long is it offline and why. Can a company like LabsLand offer an unreliable experiment deployed in a university thousands of miles away?

The installation of additional software to access and control a remote experiment is an undesirable situation that clearly compromises the security of the computer downloading it and of the network containing it (undoubtedly a university, an educational or research centre, etc.). For example, the remote-farm of Berlin University (see Fig. 2.11) requires the LabVIEW purchase plugin to be installed. Is it recommendable? Is it possible in our place of study or work? In general, and even if it were not

Figure 2.9. Example of problems of compatibility with remote experimentation.
Source: https://remotelab.fe.up.pt/.

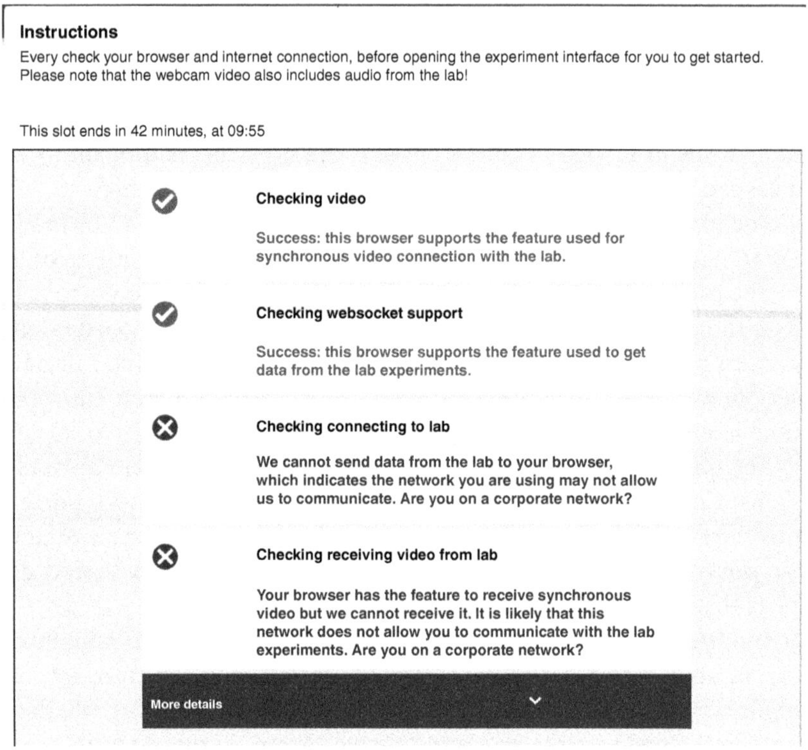

Figure 2.10. Analysis of quality of client's network in a remote experiment.
Source: http://stem.open.ac.uk/study/openstem-labs.

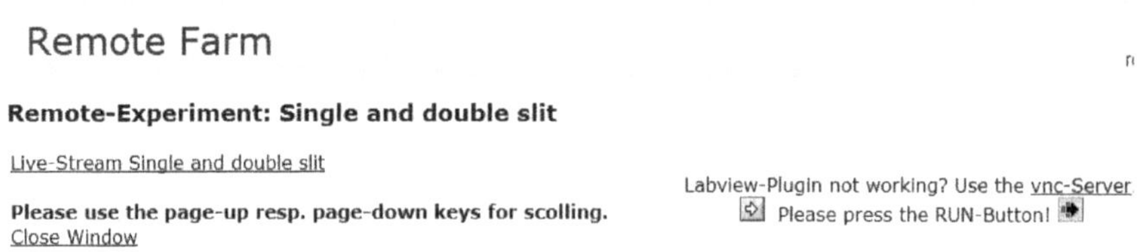

Figure 2.11. Download requirement of LabVIEW plugin in a remote experiment.
Source: https://remote.physik.tu-berlin.de.

taken into account, no employee can install software on their computer without the company's explicit consent.

For example, the extraordinary e-lab remote laboratory in Lisbon requires java 6.0 (see Fig. 2.12) to be installed, which is not only a nuisance

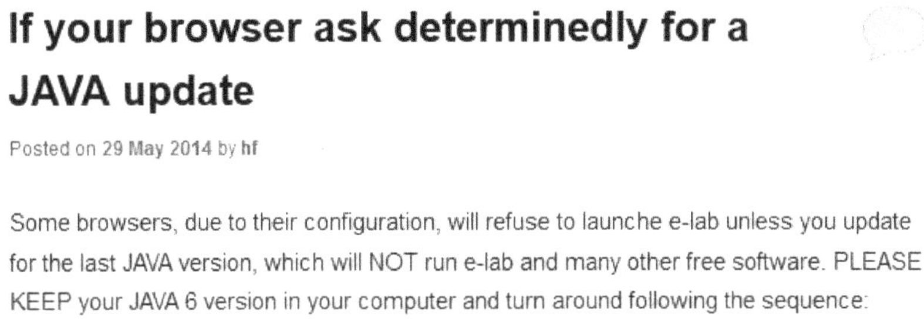

Figure 2.12. Java installation requirement for a client to perform a remote experiment.
Source: http://www.elab.ist.utl.pt/.

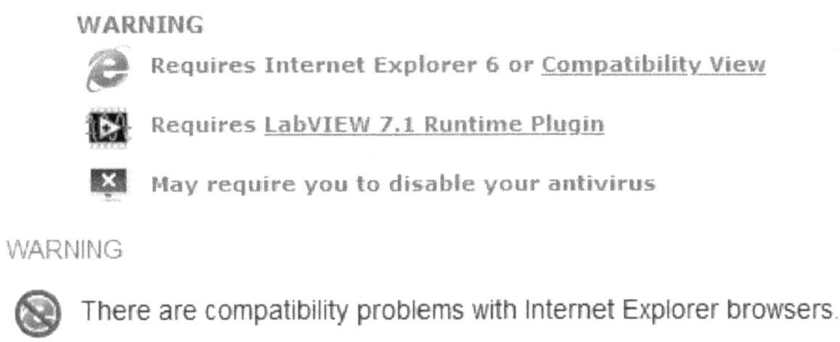

Figure 2.13. Example of incompatibility within a remote experiment.
Source: https://remotelab.fe.up.pt/.

but also compromises the universality of the laboratory and the security of the institution downloading java.

In the curious case of Fig. 2.13 we can see that in the same remote laboratory there are two remote experiments with different and almost exclusive problems.

Finally, and in relation to reliability and security, it is amusing to read two texts received from owners of remote laboratories during the tests for this book: 'It is firewall blocking the free spirit of the academics! You have a firewall problem, it is blocking ports 9000, 9001, 9002, 9003. Run in an open Internet access or ask your IT people to unblock. And have RTSP port open to see video as well (554 if I'm not wrong)', Is our educational centre's IT service going to follow such recommendations? The opposite case arises in the following email: 'Because of the recent terrorist attacks in *XXXX*, the network of the university is more strongly secured'. It is clear that the story changes considerably depending on the designer, user and IT services.

Now let us imagine that an experiment complied with the aforementioned steps and we access it, but it turns out to be occupied. What can be done? The worst solution involves 'refusing access to whoever's entering or ejecting whoever's inside'. A more efficient solution would be to manage the queue: first come, first served. But what happens if a student leaves the queue? Do they block it? What happens if a teacher finds a queue of 10 students in front of him? Well, the teacher should almost certainly go first and therefore the queue's priorities need to be managed. Queue management is far from a trivial matter.

Another criterion is using a booking system instead of a queue. In this case, the teacher or the student can book a time slot in advance in order to be able to access the remote experiment without experiencing problems with other users, but in this case the problem of underuse arises. The slots are usually 30 or 60-minutes long. What happens if after five minutes the user exits the experiment? What happens if the user who books the experiment does not access it? Do students really use the experiment for 60 minutes or do they devote this time to experimenting, noting down results or analysis? Most probably nobody else will be able to use it, leaving it underused.

There are solutions, such as the one proposed by Ferreira and Cardoso (2005), which make it possible to offer to students a time slot that has not been occupied, or the part of a slot that remains available. Or maybe the queue can be moved (emptied) as free time becomes available. There are many strategies that are easy to explain but difficult to implement.

Similarly, it is easy to think that when an educational centre incorporates a remote laboratory into its classes, it seeks to do so in a personalised manner. This means that the remote laboratory will have to be integrated within the centre's LMS, which will be accessible in the students' language, and that they will enter the latter using the usernames and passwords provided by the educational centre. This is very important not only for the students but above all because of the new European regulations with regard to data protection (General Data Protection Regulation) and their application in each country. Once again, this should be transparent for the educational centre but requires appropriate technical knowledge from the designer. Figure 2.14 shows how access to a remote robot has been personalised for an educational centre within Moodle, and the same can be done with Google Classroom or Sakai.

In another scenario, teachers are often not content with simply accessing a remote experiment to use it in class within their teaching-learning strategy but require additional information such as Learning

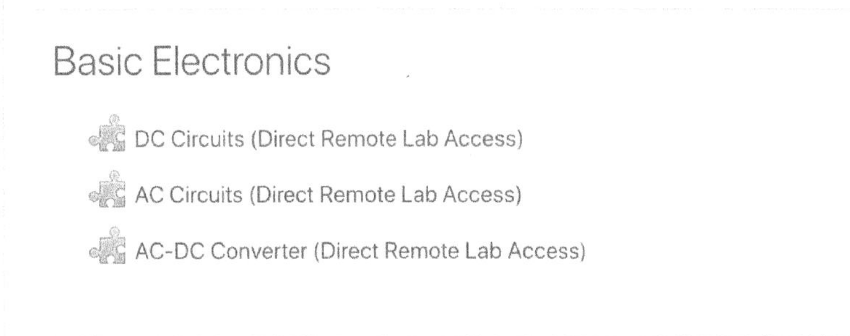

Figure 2.14. Integration of a remote experiment in Moodle.
Source: alud.deusto.es.

Analytics, LA. The use of LA helps teachers to understand better what use students make of experiments and how, as well as helping teachers complete reports on this activity. The field of LA is immense and will be addressed later in the book, but responds to basic questions like: how many times has a student accessed an experiment? How long have they been connected? What actions have been undertaken? Which programme, for instance, have they downloaded in the robot? Have they copied the programme from a classmate? These questions and others may be interesting simply in terms of the direct answers, but they also give rise to other more sophisticated questions. For example, in light of the previous answers, can teachers modify and improve their teaching-learning strategy? Can it be adapted to the remote experiment itself? Adaptive Learning based on LA is a very active and innovative research field (see Section 7.1). To be able to analyse data, the RLMS must first collect it in a database.

All the above has been based on present scenarios, what a user needs now, but it makes sense to think of the future. For example, we have two experiments and we want to add a third. In this case, a good RLMS will make the job easier and only demands a degree of effort. Even if the design is external, a good RLMS will help us.

Take a scenario where we have a programmable robot, but only one is available to use at a time. Thus, when a student is using it, nobody else can use it, and if someone wants access, they will join a queue. If the decision is taken to buy and integrate four new robots, a good RLMS should make it easy to add new copies of the same remote experiment, but technically this is very complex, particularly when we go from several users and one or two robots to thousands of users and many robots. The scalability of a remote laboratory is crucial to the development of the remote experiment.

Finally, let us imagine that our school decides to share its remote laboratory with another school, and vice versa, as a result of which both will have twice as many experiments but at the same cost. But they want to make this transparent for the student. In other words, the student can access all the remote laboratories via the web at their school. Once again, this seems obvious, but it requires the RLMS at both schools to be capable of communicating with each other so as to share the remote experiments, avoiding blockages, managing priorities, etc. The federation of remote laboratories represents the highest technical level in remote experimentation and seeks to render it more sustainable and efficient.

Returning to the beginning, for each participant in the remote experimentation, each of the four aforementioned criteria is more or less important. Universality may not be important for the designer (for whom it is sufficient that it functions), but for the student, it is. Teachers certainly will not care how difficult it is to add a new experiment (they already have their own), but the designer will. The cost will certainly interest everybody. The designer should be aware of all these requirements and should prioritise and then implement them.

To summarise, when the design of a remote laboratory addresses the aforementioned requirements — or some of them — from a professional perspective, and not as a mere aggregation of technical achievements, in other words, when the designers begin their work with a list of requirements based on the aforementioned concepts, then one is speaking of a Remote Laboratories Management System (RLMS). Compliance with these requirements is conditioned by the design technology of the remote laboratory, which in turn is the responsibility of the designer and will be addressed in Chapters 5 and 6.

2.3.2 *Requirements and features of a remote experiment*

It is obvious that the features of an RLMS will condition the experiments themselves, but the latter can also have requirements or features, independently of the RLMS in which they are contained.

When we analyse a remote experiment, we extract its features. And then these features can be used as requirements to design a new remote experiment.

An experiment can be characterised according to three major criteria:

(1) **Accessibility:** Describes how the student accesses the experiment.

(2) **Interface:** Describes how the student controls and observes the experiment.

(3) **Learning support:** Describes how the experiment supports the student's learning.

Each of these three criteria can be described in greater detail, as is done in the following:

(1) **Accessibility**

- **Universality:** This feature has already been included in the RLMS, but is almost more closely related to the remote experiment itself, as it is universality that can require the installation of plugins, open ports, use specific browsers, etc. However, as it is a feature that arises from the original design it usually affects all the experiments in a remote laboratory, and for this reason is associated with the RLMS.

- **Queues and bookings**: Access to the experiment may be via queues or bookings, or there is even a possibility that this situation is not managed. This is a feature shared between the RLMS and each experiment.

- **Interaction:** The student needs to interact online with the experiment or may simply launch the experiment.

- **Multiuser:** The experiment may only allow one student to control it (the normal case) or may allow various students to have simultaneous control.

- **Deferred in time or pre-recorded:** In this case, the remote experiment has been recorded previously and is offered to the student in a real scenario. This feature, given its importance, is described in detail in Section 2.4.

(2) **Interface**

- **Interface:** The experiment, to the extent of its abilities, should reproduce the working environment in the hands-on mode. There are interfaces that are so complicated that they literally eliminate the value of the experiment. However, it can also be improved, taking it to a higher level of interaction or observation, as is indicated by the following feature.
- **Controllability and observability:** The experiment should permit access to the greatest possible number of experiment inputs and outputs. This feature is directly connected to the read/write classification explained earlier.
- **Feedback:** Directly linked to the previous feature is the way of returning feedback directly to the user via the use of a webcam, microphone, etc. Poor, or impossible, feedback can be detrimental to the student's experience.
- **Data:** The experiment returns the experiment data via interface. It may do so as a simple number, by means of files or even processing the data in graphs. If the process is excessive, student learning may disappear.
- **Adaptability of the interface:** According to the student's level, the interface is adapted so as to increase learning results. This is an active and complex line of research as it combines technical and learning resources. See Sections 7.1 and 7.2.2.
- **Augmented reality:** Some input or output variables have augmented reality. For example, liquid is dyed to show its temperature or the lines of a magnetic field are drawn. See Section 7.2.1.
- **Virtual reality:** The user and the experiment form part of a virtual world, like *Second Life*. This effort may increase the student's sensation of immersion in the experience — not only

in the experiment — and therefore improve it. Currently, it is not a very popular technology, but it continues to be explored. See Section 7.2.1.

- **Gamification:** Integrates the remote experiment as part of a game experience, where optimum control of the experiment represents an advantage in the game for the player/student.

(3) **Learning support and instructional design**

- **Didactic support:** The experiment may be accompanied by a file that explains how it fits into the curriculum. Its absence may not attract teachers to the experiment, but its excess may drive them away if they regard themselves as 'below' the didactic material. This aspect is addressed in detail in Section 4.2.

- **Dashboard and learning analytics:** Once the RLMS collects the data of the user's activity, analysis of the latter depends on each experiment, so there will be a Dashboard for each experiment. Use of the Dashboard will enable teachers to draw conclusions with which to improve their subject or evaluate their students. This analysis may also form part of the adaptive interface seen earlier. It is an active and complex field that is described in more detail in Section 7.1.

- **Online tutorial and learning analytics**: The Dashboard analyses the data *a posteriori*, but it is also possible to track students' activity in real time and, depending on what they are doing, the tutorial or the chatbot could advise students how to correct mistakes they are making or recommend new activities to improve their learning. This is an active and complex field that is described in more detail in the NMC Report (2019).

The quality of the experiment — which depends on the experiment itself —, the sensation of immersion therein and control of the experiment and its data are the main features of an experiment. These can be improved with the correct use of technology, but priority should always be given to control by the student and the learning experience. The most characteristic element is the interface, and the term usability summarises all the aforementioned features.

Earlier, we described the various types of remote experiment (in Section 2.1) and now we will describe their features. Both sections are related but are different. The types refer to the experiment itself and to how it is presented to the student, in other words, to the experiment before it is performed, while most of the aforementioned requirements or features are related to the implementation of the design and to the student's actual use of the remote experiment. It is easy to connect both sections by means of a table, conceptual map, etc.

Let us consider some examples of remote experiments and their behaviour in relation to the aforementioned features. The control of the robot offered by the University of Western Australia (Telerobot) requires a plugin (see Fig. 2.15) to be downloaded, which is far from recommendable and may even be prohibited by the educational centre.

Meanwhile, some experiments have highly complex interfaces, which do not facilitate the student's work or their learning. Figure 2.16 shows the

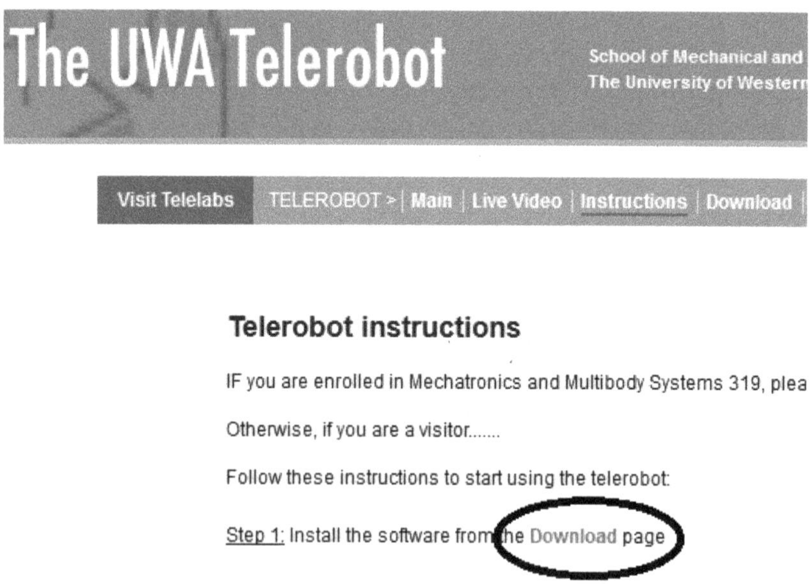

Figure 2.15. Download requirement of an unknown program.
Source: https://telerobot.mech.uwa.edu.au/.

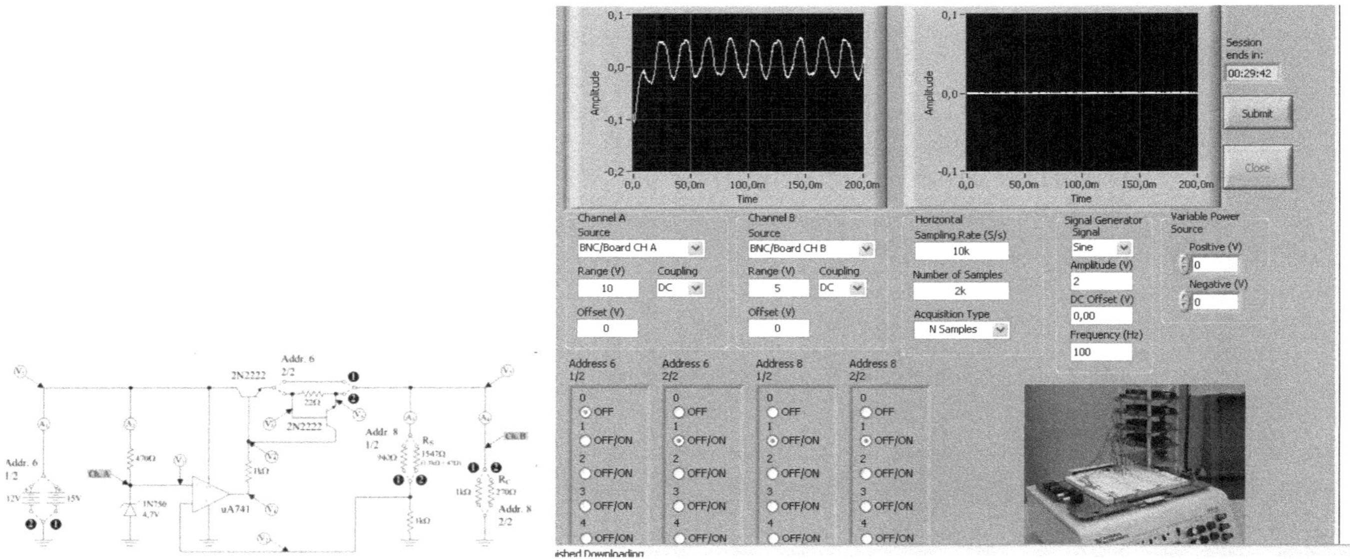

Figure 2.16. Interface of an electronic circuits remote experiment designed by ISEP (Porto, Portugal).

Source: Sousa *et al.*, 2010.

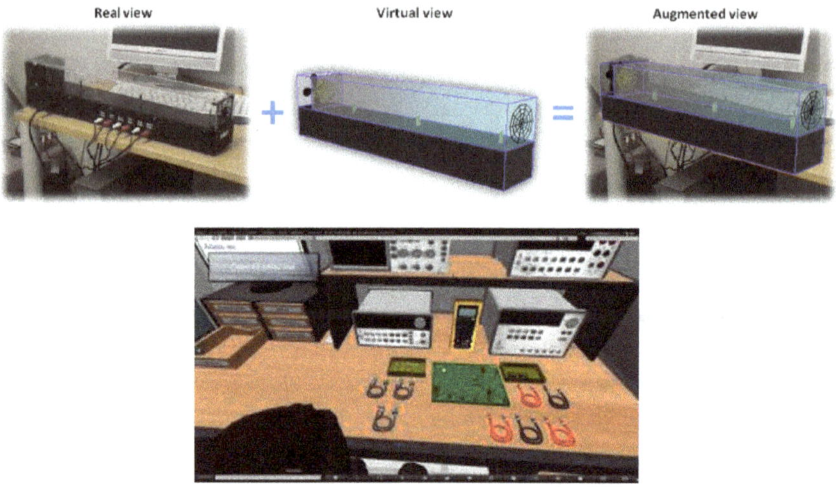

Figure 2.17. Examples of interfaces of remote experiments with augmented and virtual reality.

Source: https://unilabs.dia.uned.es/and https://3dlabs.upm.es/.

interface for assembling quite simple circuits. Does it resemble the user's normal work? Will the student really learn with this interface?

In the opposite case of a poor interface are those that improve the usual interface, for example via the use of augmented reality or the use of virtual environments. The UNED's UNILabs laboratory (Spain) integrates elements of augmented reality in its control experiments, while UPM's eLab3D remote laboratory (Spain), for experiments with electronic circuits, uses virtual reality technologies (see Fig. 2.17).

Callaghan (2012) goes one step further by integrating some of the aforementioned aspects into a game the central element of which is an electronic remote experiment (see Fig. 2.18). To progress in the game, the student must overcome certain challenges related to electronics, all under the aesthetics and dynamics of a game. Gamified remote laboratories and games with remote laboratories have a clear and innovative future within serious games.

Feedback and data processing are fundamental elements of an experiment. On the left of Fig. 2.19, we can see that the experiment data cannot be downloaded from the RexLab interface, but must be copied by hand by the student on an excel document or similar. On the right of

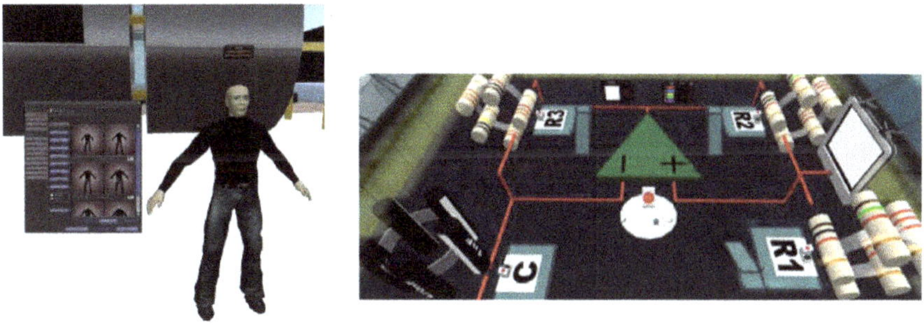

Figure 2.18. Computer game based on remote experimentation.
Source: circuitwarz.com.

Fig. 2.19, MIT not only allows for data to be downloaded but also to be visualised and for graphics to be configured.

The aforementioned features can be divided into two groups: technical and didactic. The former has more to do with the designer and the RLMS, while in the latter teacher and students are protagonists. The former can conceptualised as trustability, and the latter as usability. Often it is not possible to achieve both, or worse still, sometimes a high degree of trustability can mean a reduction in usability.

In Chapter 3, different RLMS and remote experiments will be analysed according to the criteria described. To this end, it is important to study in greater detail the trustability and usability of experiments that are deferred in time because they are pre-recorded.

2.4 Ultraconcurrent, Pre-recorded or Time-Deferred and Data Set Remote Experiments

A common feature of pre-recorded and Data Set remote experiments is that the student does not access them in real time but in a deferred fashion. In other words, they may not be considered remote experiments, but they are at least for the research teams that work with them. For instance, the GOLC (Global Online Laboratories Consortium) association awards an annual prize for remote labs and both types of remote experimentation compete.

In a Data Set experiment, the educational experience is focused on the data obtained from the experiment, not on the way in which it is obtained,

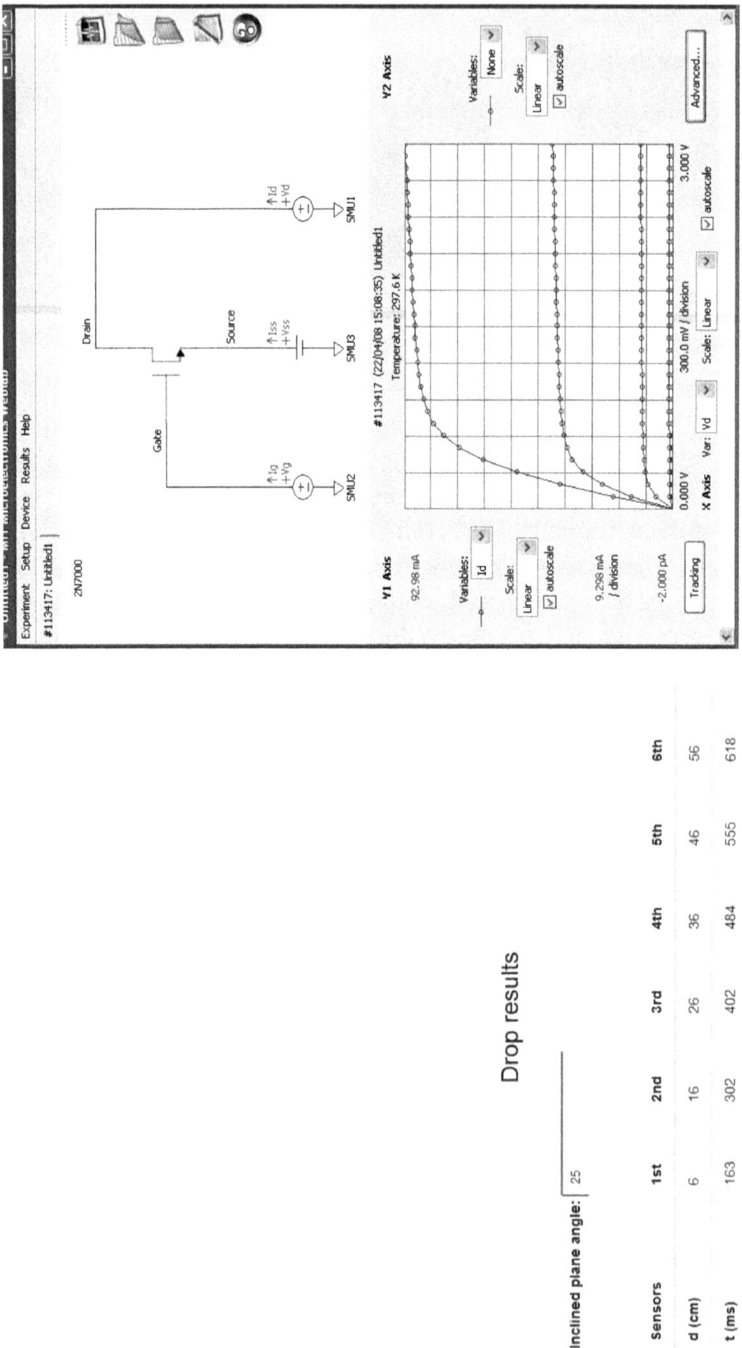

Figure 2.19. Interfaces of remote experiments with visualisation of data in iLAB (MIT, USA).

while in a pre-recorded experiment what is important is the student's experience, and not whether this is online or deferred.

They appear to be dissimilar approaches, but they have elements in common, and they are presented in detail in the following.

2.4.1 *Data Set*

Let us imagine an experiment that seeks to generate particle collision in LHC (Large Hadron Collider) of the CERN to analyse the results of the trajectories of the collision. A student (in this case at university or in the final year of secondary education) could remotely access the LHC (see Fig. 2.20) and generate this collision. Would it make sense to do this or is the important thing to analyse the results of some previous collision? Students may reinforce their knowledge if they themselves produce the collision, but what is most important is analysis of the trajectories recorded and available in a Data Set. Clearly, it does not cost the same to generate collisions at the user's request as it does to use data already stored.

Think of astronomy and the Sun: is it necessary for the student to control a telescope and take photos of the Sun or is it sufficient to analyse them in detail to determine the length of the solar year? To begin, the student should take photos over a period of many days. Is this process and its duration relevant to the student's learning? Almost certainly not, although the instructor will have to make this decision.

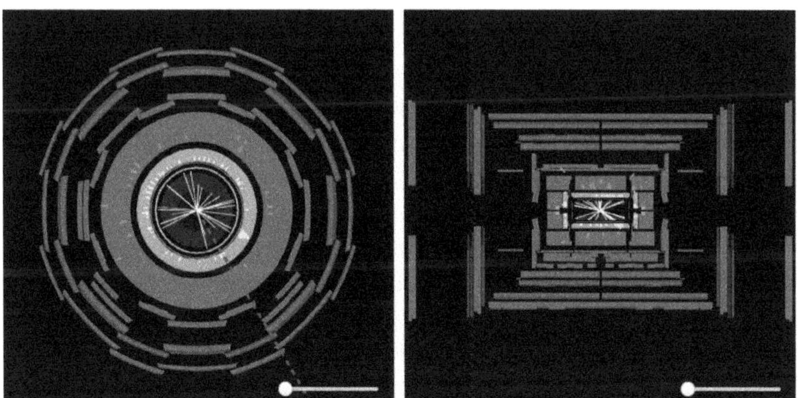

Figure 2.20. Example of Data Set type remote experiment in GoLab.
Source: hypatia lab in https://www.golabz.eu/.

A somewhat more profound analysis of what a Data Set is would certainly lead us not to regard it as a remote experiment, as the user does not interact at all, but simply sees the results, which might as well be written in a book. If the Data Set experiments are included here, it is because academic literature usually includes them alongside remote experiments, and for this reason the reader should recognise them.

2.4.2 *Time-deferred, pre-recorded or ultraconcurrent remote experiments*

Let us consider Archimedes' laboratory in more detail: the student launches objects into test tubes and sees whether or not they float via webcam and also sees the values of the output variables. The student can only do two things (combinational experiment): raise and lower the ball (see Fig. 2.7); the rest is observation. In this case, the designer can record on video the rise and fall of a ball a considerable number of times, so that when a student launches a ball into the water, they may be watching a launch online or one recorded earlier. The experience is the same, as the recording is via interface, and it is the instructor or designer who decides whether or not to inform the student of this fact (that the experiment is pre-recorded).

The same thing can be done with an inclined plane: the student places the ball on a plane with a certain gradient, releases it, watches its movement and observes the output variable. Is it possible to record the fall of the ball for every possible departure angle and then offer these recordings on a deferred basis?

Can the same thing be done with a pendulum and different launch angles?

If what is recorded is the actual interface, is there a way of differentiating between what has been recorded and what happens in real-time? Is the data not the same and therefore the same analysis possible? Is there learning? Are these real doubts or preconceptions? In this scenario, it is worth bearing in mind that students are already digital natives and that for them reality is not divided into digital and real — both are real for them. Instructors must think about whether their own ways of working contain bias and whether this takes them further away from or closer to the students' way of learning.

At the other extreme is the mobile robot, as the student will create a programme and execute it in the robot. Can this behaviour be recorded and offered on a deferred basis? Obviously not, as it seems impossible for the designer to be able to imagine and create a set of all the possible programmes that their students might create. The automaton that the student is going to design will, by definition, be finite (a microprocessor is a deterministic finite automaton), but the whole set of possible automata is so enormous as to be infinite as far as the designer's possibilities are concerned.

Is the online/batch feature of an experiment seen in Section 2.1.2 related to the possibility of deferred experiments? It may seem that a batch experiment can be offered on a deferred basis while an online one cannot, but this is not necessarily the case.

Batch means that the student launches the experiment now and receives the result later, with no interaction during the experiment. The result will be obtained for each experiment and may be delivered via a file, a graph or a video to be sent to the student by e-mail, for instance.

Time deferred means that when the student launches an experiment, the latter will have been recorded and will be in a database, so there will be no real execution of the experiment, but reproduction of an experiment and its pre-recorded interfaces.

For example, the robot programmed by the student to follow a line may be batch, but under no circumstances can be offered on a deferred basis, as the designer cannot respond to all the potential different algorithms. However, his algorithm cannot be created in its totality by the student, but only parametrised in one of its aspects. In this case, the different algorithms are finite and the designer can launch and record them *a priori* to offer them later on a deferred basis.

In general, a deferred or pre-recorded experiment is usually online with limited interactivity (it parametrises and launches the experiment), so the student has the sensation has the experiment is real, and not a recording. It has already been mentioned that what is recorded is not the experiment in a hands-on laboratory, but the entire interface.

However, it is true that an online experiment where the student has to interact during the latter cannot be offered on a deferred basis, or if it can, this is not of interest or is extremely complex. It is virtually impossible for the designer to predict and record the interaction during the experiment.

Where a relationship does exist is between combinational and sequential experiments and their possible deferred mode. If an experiment is combinational, then it may be offered on a deferred basis.

In this case, it is important to analyse all the possible input situations. For example, if there are 10 objects in the Archimedes Experiment, then it will be sufficient to raise and lower them (2 actions × 10 objects = 20 events). If this repeated 20 times per object, we will have to manage a total of 400 events to be recorded. Every time the student requires an object to be lowered, the remote experiment will search in its databases for one of the 20 videos of lowerings and show it to the student on the interface.

In the case of the inclined plane, we are talking about 90 angles, and if we repeat each angle 20 times, we have to record 1800 events during the experiment. In the radioactivity experiment, there are around 10 000 events. Is it possible to record, store, handle and show the user so many events? How long does it take to record 10 000 different situations? How much space do they occupy?

A fundamental feature of time-deferred remote experiments is that several students can access them at the same time, as they are not real experiments, but recordings. This is why these remote experiments are also given the name Ultraconcurrent. The absence of queues of students waiting to access an experiment is a very valuable feature of this type of experiment and affects its trustability (see Section 2.5).

Consider another feature: Can a time-deferred remote experiment be blocked or broken? Will it have maintenance issues? Will it be necessary to check the status of the motors that raise the pendulum or the inclined plane?

It should be emphasised that every Data Set experiment is deferred in time. The difference between both lies in the fact that in the deferred experiment users have a sensation of control and interactivity, while in the Data Set experiment they simply evaluate the data downloaded. In the deferred model, the experiment is recorded on the interface, while in the Data Set version the data is submitted on an analysis interface. In the deferred experiment, the user's experience of control of the experiment is crucial to the user's learning, while in the Data Set experiment the tools used to analyse the complete set of data provided is crucial to the learning experience.

Let us see how the two experiences are related. For example, if an electronically sensored pendulum exists (not necessarily remotised), the designer can launch it 100 times in order to measure and store in a file its angular position at each moment. Once this has been done, the student neither accesses the pendulum, nor launches it nor sees its evolution, but simply receives an excel file with thousands of data that must be analysed so as to draw the pertinent conclusions (Data Set). In this case, what is important is not the experience or seeing the pendulum, but the data. In the deferred experiment, the student does not access the data directly but

Table 2.2. Analysis of types of experiments.

Type of phenomenon	Type of experiment	Type of control	According to the designer	According to the user
Real	Read	Without control	Batch	Batch
Simulated	Parametrised	With control	Online	Passive
	Logical			Interactive

recreates the way (the interface) in which it is obtained. Whether or not one approach is preferable to the other is up to the instructor.

Let us return to the type of remote experiments in Section 2.1. We have already said that, depending on the experiment to be remotised, this will affect its implementation. Table 2.2 presents a clear of example of remote experiments deferred in time, showing in yellow those that do not prevent it in theory but do so in practice. Marked in green are the types that allow for the design of a pre-recorded experiment.

2.5 Trustability and Usability

As has been indicated, trustability and usability are different concepts or criteria that must be analysed in a different way, but both condition the educational value of the remote experiment, and therefore it makes sense to study them together.

In the previous section, we note that the technical aspects that condition the trustability of the remote laboratory tend to be objective. Thus, in order to know whether a remote laboratory is universal one simply has to access it via any kind of device with any browser under any OS. To know whether it is reliable one simply has to access it. However, a detailed study of these features poses a dilemma between universality and power or efficiency; the latter understood as the cost involved in providing the laboratory for remote experimentation. For example, if one wants the remote laboratory to consume little bandwidth (a problem in some schools and countries), this often involves employing non-standard solutions that compromise security, or if one wishes to integrate a very powerful module, for instance a LabVIEW instrument panel, this decision will inevitably require the use of a plugin. Or if one wants a very eye-catching interface, this will condition the universality of the laboratory.

In other words, in quite a few cases a more powerful or efficient remote laboratory will involve the impairment of its universality and, with this, its trustability. This dichotomy is not easily resolved. This question was put to a group of nine researchers, all of whom said that universality was the most important feature and that an increase in the power of the remote laboratory could not justify a loss of universality (García-Zubía *et al.*, 2009a). Table 2.3 shows the opinion of each researcher: the lower the value assigned to each feature, the greater the importance attributed to the latter. The first three features of the table constitute universality.

It is worth noting that, nowadays, technological progress is focusing on combining both characteristics. For example, at the start of the 21st century many remote laboratories were web applications downloadable on the desktop, but now all are web applications that strive to avoid the use of plugins. Similarly, in the design of interfaces, the HTML5 standard was initially not very powerful and conditioned the processing speed and the bandwidth of the remote laboratory as a whole. The alternative to this situation was to use ad hoc solutions, which limited the standard nature of the laboratory. Now, however, HTML5 and similar technologies like web sockets barely condition the interfaces of remote laboratories.

Table 2.3. Opinion of experts in remote experimentation (Garcia-Zubía *et al.*, 2009a).

	E1	E2	E3	E4	E5	E6	E7	E8	Total
Cross platform	2	4	3	1	2	1	1	1	15
Security	1	2	5	3	5	2	2	3	23
Web browsers	3	5	4	2	3	3	4	2	26
Intrusivity	4	1	8	4	1	10	5	5	38
Interaction	6	7	2	7	6	4	3	6	41
Installation	5	3	9	6	4	5	7	4	43
Devices	10	6	10	5	7	9	6	7	60
Bandwidth	8	9	1	8	9	7	9	10	61
Audio & Video	9	8	6	9	8	6	8	8	62
Power	7	10	7	10	10	8	10	9	71

Notes: E1: IT service of U. Deusto (Spain); E2: Javier García-Zubía, U. Deusto; E3: Ingvar Gustavsson, BTH (Sweden); E4: Dieter Müller, U. Bremen (Germany); E5: Gustavo Alves, ISEP (Portugal); E6: Andrea Bagnasco, U. Genoa (Italy); E7: Jesús A. del Álamo, MIT (USA); E8: Denis Gillet, EPFL (Switzerland).

Similarly, and in a specific fashion, using LabVIEW Virtual Panels boosted and greatly simplified the design of remote laboratories, but significantly conditioned their universality (plugin, exclusive use of Internet Explorer, open ports, etc.). Today, however, the situation is very different, and the LabVIEW Virtual Panels are based on HTML5 and JavaScript.

It is important to highlight the fact that although technology has advanced in terms of power and standardisation, there has been a corresponding increase in the demands made on the designer of remote laboratories. The design team must have a high level of IT technical skills, and not only regarding the experiment in question but to guarantee the trustability and power of the remote laboratory.

Having ensured the trustability of a remote laboratory, it is essential to maximise its usability. Usability is basically conditioned by two elements: the interface and the experiment's educational content. The latter is not of a technical nature, is addressed later and depends on the teacher and the curriculum.

From a technical point of view, usability is a feature that designers cannot measure *a priori*, they may sense that it is going to be high — that the remote experiment is going to be usable. Usability should be measured with the user as protagonist. In some cases, the designer prepares a qualitative survey in which users are asked about their impressions and are requested to score each question or aspect, generally on a Likert scale of 5 or 7 points. This kind of questionnaire is easy to make and understand, but the results of different remote experiments are difficult to compare as the questionnaires are different. These kinds of questionnaires are usually prepared by the designer or the instructor, which conditions their credibility since it is difficult to accept that the person who designs an experience is also responsible for its evaluation without indulging in self-fulfilling prophecy. This aspect is studied in detail in Section 7.3.

Very useful in order to analyse the above is Lahoud and Krichen (2010), as it combines the elements described in this section. The analysis in Fig. 2.21 asks users to choose between Fidelity, Accessibility and Usability.

- Fidelity is understood as the remote experiment's capacity to resemble the original hands-on one.
- Accessibility expresses the fact that a remote experiment is accessible at any time and place from any computer terminal.

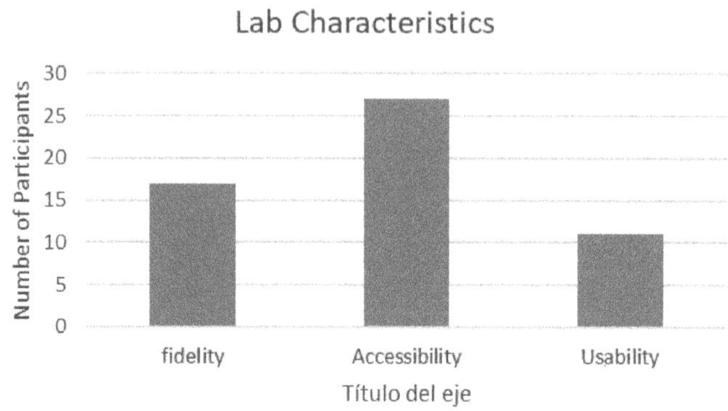

Figure 2.21. Students' opinion regarding Fidelity, Accessibility and Usability (Lahoud and Krichen, 2010).

- Usability is the remote experiment's capacity to be controlled by the user.

The results shown in the image highlight that the property most highly valued by students is Accessibility; in other words, being truly able to access the remote experiment. In second place in terms of importance, respondents place fidelity. And the least important aspect is the usability of the remote experiment. These results are in line with those obtained by Garcia-Zubía *et al.* (2009).

In the classification included in Section 2.1.2, the time-deferred remote experiment was introduced. This kind of experiment — psychological considerations apart — guarantees accessibility as it is never broken and never creates queues of students, therefore respecting the main requirements according to Lahoud and García-Zubía. Fidelity and Usability basically depend upon the experiment's interface, and not on the underlying technology. In this sense, the design and use of ultraconcurrent pre-recorded remote experiments is an option to be considered whenever possible.

Part
2

Examples of Remote Experiments and Pedagogical Value

Chapter 3

Use Cases

This chapter provides a detailed description of different remote laboratories and remote experiments under the characteristics and criteria described in Chapter 2.

3.1 Examples of Remote Labs and RLMS

A remote laboratory is distinguished by two aspects. On the one hand, a remote laboratory offers the user various remote experiments from, if possible, different knowledge areas. On the other, a remote laboratory has an RLMS that assigns common characteristics to all experiments.

Below is a description of the general characteristics of several remote laboratories regarded as exemplary. Later, in Section 3.2, the reader will find detailed examples of the experiments, as this section only addresses the general characteristics provided by the RLMS.

UNILabs

Bibliographical analysis (Heradio *et al.*, 2016a) revealed that Sebastián Dormido's group in the UNED (Spain) is the most active. Its remote laboratory is UNILabs (https://unilabs.dia.uned.es/), and it offers over 24 remote experiments, mostly in the area of control engineering. All the experiments have didactic support and are integrated in Moodle, and to access most of them a username/password is required. UNILabs does not require the installation of applications or compromise user security, and its design facilitates the scalability and extensibility of the laboratory. Many of its remote experiments are reinforced via virtual and augmented reality. It is a perfect reference of the state-of-the-art in remote experimentation, and the only criticism is that it focuses on one single type of remote experiment (automatic control systems) for the university.

NCSLab, in China, is a remote laboratory and it is similar to UNILabs in areas of interest and use of virtual and augmented reality (www.powersim.whu.edu.cn/ncslab).

WebLab-Deusto

The RLMS WebLab-Deusto is an open source (https://github.com/weblabdeusto) platform developed by the University and by LabsLand (https://labsland.com/es/labs) and has been deployed in different universities around the world. To date, the RLMS WebLab-Deusto has been deployed in more than 15 universities in nine countries from four continents and has served over 200,000 sessions for over 10,000 users over 15 years.

WebLab-Deusto and LabsLand have the same characteristics as UNILabs but are also the only remote laboratories with federated experiments in different countries (South Africa, Costa Rica, Germany and the USA). Their experiments are very diverse in terms of subject matter and audience — robots, electronics, physics, biology, etc. — and can be integrated within any LMS platform. All their experiments are accessible without any kind of technological restriction, but require a username/password or direct contact with the company. Their main characteristic are the professionalism of the service and the companies' commercial and collaborative interests (placing an experiment in the federation in exchange for having access to others). Worthy of mention is the fact that the companies have both designed a method to develop ultraconcurrent Data Set or pre-recorded experiments that are unique the world over.

RexLab

The RexLab platform (https://rexlab.ufsc.br/) has been developed by UFSC in Brazil and has many technological and didactic characteristics in common with the two aforementioned laboratories. It resembles LabsLand in its scientific variety, and in fact LabsLand shares with RexLab some ultraconcurrent experiments. The main problem with RexLab is reliability, as its experiments are often unavailable due to connection problems and the like — Problems that are not attributable to the research team, but which clearly hinder its usefulness in the classroom. RexLab has won various awards, enjoys international recognition, and collaborates on an institutional basis with schools from all over the country in order to improve their pupils' scientific level.

ISES (Internet School Experimental System)

The ISES (http://www.ises.info/index.php/en) and RemLabNet (http://www.remlabnet.eu/) remote laboratories have been developed in the Czech Republic and Slovakia by Dr. Lustig and Dr. Schauer, respectively, and both use ISES remote experimentation hardware. Both offer physics and some chemistry experiments mainly, but none involving programming or similar. Both laboratories are documented, although they are not always available due to different problems, which, as in the case of RexLab, hinders their use in the classroom. Noteworthy in the case of RemLabNet is the fact that it is deployed in the cloud and that the experiments feature a mechanism to indicate to the user whether or not they are available, by means of a traffic light. Its main advantage is the variety and scientific quality of the physical experiments, while its main disadvantages are that of the 20 remote experiments offered, only a few are available (out of 20 at RemLabNet, only five were accessible) and that its interface could be improved. ISES and RemLabNet are two very important laboratories that require some technological adaptation.

Remote Farm

The UTB in Berlin (Germany) offers the Remote Farm (https://remote.physik.tu-berlin.de/en/). This remote laboratory mainly offers science experiments based on physics. Experiments that are accessible via username/password in unsafe mode, are complex and sophisticated, and are documented, but require the installation of a LabVIEW plugin and connection via VNC. Neither of these aspects facilitates use in the classroom, though the laboratory claims to be adapting its experiments to new technologies. This remote laboratory is in perfect condition and underwent improvements in 2019. Unfortunately, however, its video streaming is easily blocked by the educational centre server in some browsers.

Remotely Controlled Laboratories

At Munich University (Germany), the Remotely Controlled Laboratories RCL (http://rcl-munich.informatik.unibw-muenchen.de/) remote laboratory offers a set of interesting and complex remote experiments orientated towards basic physics without the need for a username/password. However, and as they themselves comment, 'The maintenance of hardware, software and especially the security adjustments to the new operating systems are now extremely complicated due to the design of the project on Windows XP. Due to the hardware, this is no longer possible for all experiments.' In other words, this well-known remote laboratory has become technologically obsolete and is no longer available.

VirtualRemoteLab

Also in Munich, the Ludwig Maximilian University of Munich, LMU, offers the VirtualRemoteLab remote laboratory with a very specific set of remote physics experiments using a spectrometer (https://www.didaktik.physik.uni-muenchen.de/remotelab/index.php?lang=en&page=1). The experiments are accessible without username/password and are accompanied by didactic documentation. From the software perspective — security, universality, etc. — it is a perfect laboratory, in which it is worth noting that several remote experiments are offered in pre-recorded formats and are therefore ultraconcurrent. This is a good example of a recently created remote laboratory.

GOLDi

In Germany, the University of Ilmenau has designed GOLDi (Grid of Online Laboratory Devices Ilmenau) that offers a set of remote experiments orientated towards programmable systems: microcontrollers, FPGA and PLC (https://www.goldi-labs.net/). GOLDi is the reference remote laboratory in the field of programmable systems and offers its university students the possibility of controlling more than 10 sophisticated electromechanical mock-ups, some of which are installed in other countries. The student needs a username/password and access is not secure, but all the development tools are included in the environment. GOLDi needs technological adaptation to new web standards in order to guarantee its universality.

Online Experimenta FEUP

A very special remote laboratory is Online Experimenta OE@FEUP (https://remotelab.fe.up.pt/), backed by FEUP (Portugal), directed by Teresa Restivo, and the winner of many awards. Upon accessing the site, the user sees that some remote experiments are HTML, and therefore easily accessible, while others are still based on LabVIEW v7 and thus need a plugin, which is a recurring problem in remote experimentation. What makes this laboratory unique is the intensive use of virtual and augmented reality to complement the remote experiments, as well as the use of haptic devices. The student, therefore, has the remote real version and the remote virtual version with AR/VR, which makes the learning experience richer and more complete. Its experiments have didactic support in Moodle, require username/password and booking, and are focussed above all on the characterisation of devices, in mechanics and in material technology.

e-lab

Also in Portugal, e-lab (http://www.elab.ist.utl.pt/), the remote laboratory directed by Horacio Fernandes of the IST in Lisbon, offers a multitude of basic science remote experiments, especially in physics and chemistry. All the experiments are accompanied by thorough documentation and a superb mathematical approach. The remote experiments are sophisticated and complex. For example, e-lab offers a global pendulum in order to measure its oscillation at different latitudes of the world. However, this remote laboratory requires the installation of the Java plugin and the opening of a number of ports, which makes it almost impossible to use in class for reasons of security. In their day, the Java applications were perfect, but no longer. The situation of e-lab, like that of OE@FEUP, poses an important question once again: who supports a remote laboratory?

The Open University Remote Labs

The Open University in the UK offers as part of its distance training a set of remote and virtual experiments accessible via username/password (https://learn5.open.ac.uk/course/view.php?id=2). Its technology is diverse and different for each remote experiment, so some present problems while others do not. The offer includes didactic support and ranges from robots to microscopes. The RLMS does not exist and its role is played by the e-learning platform itself. Its main disadvantage are the aforementioned technological diversity and a certain technological obsolescence, since some experiments use Adobe Flash, which is no longer accepted by most web browsers.

iLABS Stanford

In the USA, Stanford University offers iLABS (http://ilabs.education/listing), a set of six recorded or ultraconcurrent physics experiments, (Neustock *et al.*, 2017) (Torre *et al.*, 2020), which can be accessed without username/password in unsafe mode in not all browsers (at least not in Internet Explorer).

iLAB MIT and LabShare UTS

iLAB and LabShare were excellent examples of remote laboratories directed by Judson Harward and David Lowe at MIT (USA) and UTS (Australia), respectively, and in their day were a world reference due to both the sophistication of their experiments and the quality of their RLMS, which permitted scalability, universality and federation from their core. However, both are more or less inactive, and neither can be used in class in a secure fashion. Both projects benefitted from generous funding (Microsoft at the MIT and the Australian government at UTS) with an initial focus based on rigorous reports, but they became unsustainable. The same thing occurred with the excellent Telelab, designed by Marco Cassini at the University of Siena (Italy).

NCSLab

NCSLab, in China, is a remote laboratory is similar to UNILabs in areas of interest and use of virtual and augmented reality. The link is www.powersim.whu.edu.cn/ncslab.

FarLabs

FarLabs (Freely Accessible Remote Laboratories, https://www.farlabs.edu.au/) is a new network of remote laboratories coordinated by La Trobe University in Australia. Its remote experiments focus on radioactivity, the environment and physics, but the majority are not yet available and pose online security problems (open ports and firewalls, webcam, no https, etc.). All the remote experiments require usernames/passwords for their effective control and belong to Explore, from the Engage-Explore-Explain-Elaborate-Evaluate learning process.

Table 3.1 offers a summary of the above. The evaluations are open to debate, which is why each reader is encouraged to make their own assessment of the laboratories presented here or of others.

Table 3.1. Analysis and classification of remote laboratories.

	Area	Security	Install	Reliability	All Web	User/Pass	LMS (1)	Fed	Final
UNILabs	Control Eng.	High	N	High	Y	Y	Moodle	N	High
WebLab-Deusto LabsLand	All	High	N	High	Y	Y	All	Y	High
RexLab	All	High	N	Low	Y	N	Unknown	N	Med. – High
ISES & RemLabNet	Physics	Low	N	Low	Y	N/Y	N	N	Med. – Low
VirtualRemoteLab	Physics	High	N	High	Y	N	N	N	Med. – High
GOLDi	Progr. Devices	Low	Y	High	Y	Y	N	N	Med. – High
OE@FEUP HTML/LabVIEW	Mechanics	High/Low	N/Y	Medium	Y/N	Y	Y	N	Med.
e-lab IST	Physics	Low	Y	Medium	N	Y	Unknown	N	Med. – Low
Stanford iLABS	Physics	Low	N	High	Y	N	N	N	Med.
iLAB & LabShare	Physics & Automation	High	N	High	Y	Y	N	Y	High
FarLabs	Physics	Low	N	Medium	Y	Y	N	N	Med.

Note: User/pass means that the student has to create a specific account in the RLMS, while LMS means that username/password is the one the student has in Moodle or in the educational centre, and it is not necessary to create accounts.

3.2 Examples of Remote Experiments

Having categorised remote experiments, it is time to present specific examples and analyse these according to the criteria described in Section 2.3.2.

Hundreds of remote experiments are available on the Internet, although for different reasons not all are effective. The following description offers the reader a diverse and accessible selection of current remote experiments, including those offered by the LabsLand company and other classics from the current state of the art.

3.2.1 *LabsLand remote experiments*[1]

LabsLand is a company created in Spain in 2015 as a spin-off of WebLab-Deusto of the University of Deusto, and it offers the user a set of remote experiments with a diversity of subject matter, objectives and technology.

At the time of this writing, LabsLand provides over 30 remote experiments at 24 universities in 14 countries all over the world. Thousands of students in different countries use it on a regular basis.

In Section 4.2, some of these experiments are described in more detail from a didactic perspective.

Time-Deferred or Ultracurrent Experiments

LabsLand offers nine deferred experiments: Archimedes, inclined plane, pendulum, planarians, radioactivity, AC circuits, spring, optics, chemistry, electronics and microscope. In all of them, the designers have recorded thousands of interactions that are subsequently offered to the reader. In other words, when the students perform an experiment, what they see is not the experiment in real time, but recorded. These experiments have been described in detail in Section 2.4.

All these experiments are accessible and universal, and because they are deferred are multiuser, it means that the student never has to queue to access the experiment.

The interface of each experiment is graphic and powerful, but under no circumstances offers the student experiences of augmented or virtual reality. Neither do they include online tutoring or any system of online support.

---

[1] https://labsland.com/en/labs.

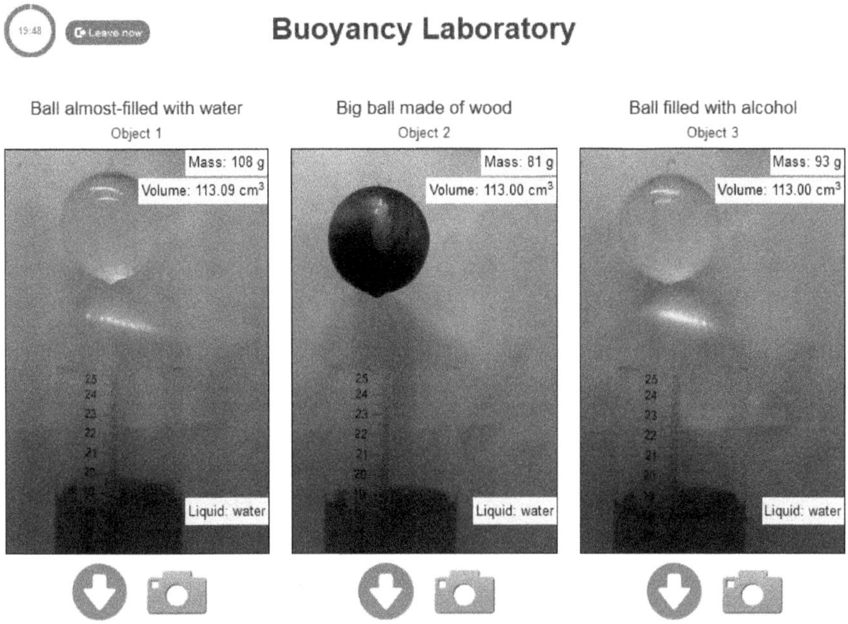

Figure 3.1. LabsLand Archimedes' Principle remote experiment.
Source: www.labsland.com.

All of them function in a similar fashion: the student sets up the experiment assigning a value to the input variables, initiates it and observes the result. The results are offered in graphic or numerical form, depending on the experiment.

All the students' activity is recorded and offered to the instructor for the corresponding analysis. In other words, LabsLand offers the instructor Learning Analytics tools.

Each experiment is briefly described below, although Section 4.2 includes deeper didactic analysis.

Archimedes

The student chooses the object to launch into the test tube to see whether or not it floats. The student makes a note of the characteristics of each object and each liquid for subsequent analysis of their behaviour (see Fig. 3.1).

This experiment does not offer numerical feedback. In other words, the student must record the information on a document, spreadsheet, etc.

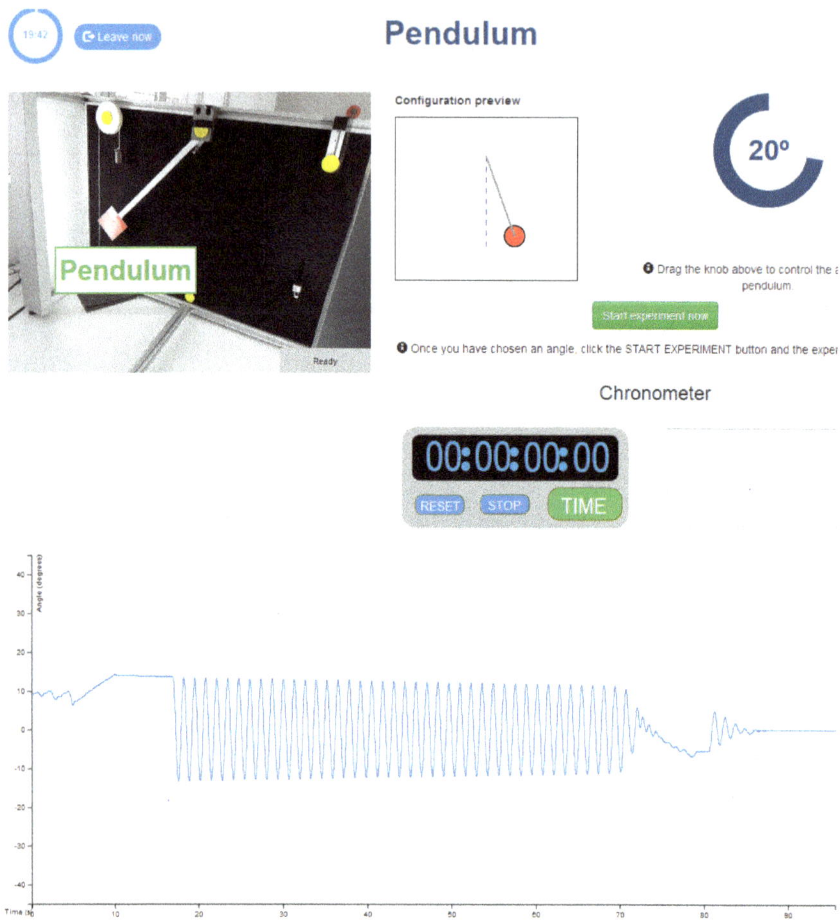

Figure 3.2. LabsLand pendulum remote experiment.

Source: www.labsland.com.

All the important buoyancy variables can be controlled by the student: mass, volume and liquid.

Pendulum

The student lifts the pendulum to a specific angle and releases it to see how it oscillates and measure for how long, depending on the angle of release (see Fig. 3.2).

The interface obliges the student to measure the period of time using the chronometer, the graph or the downloadable data file.

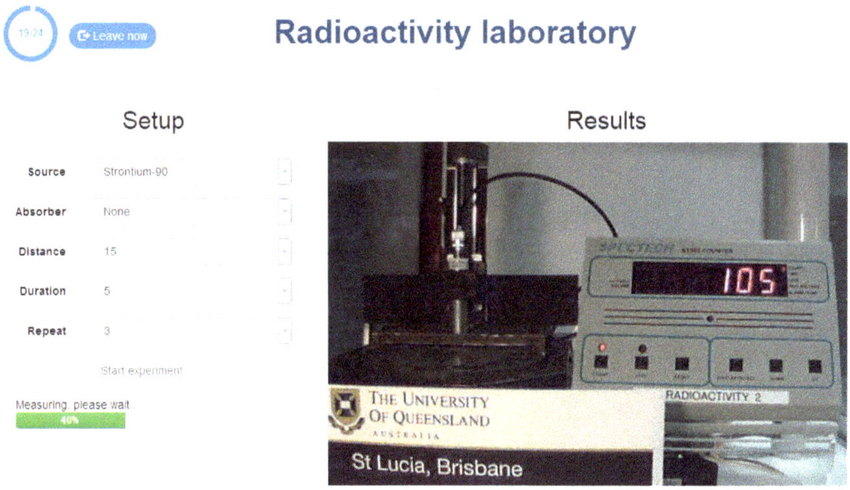

Figure 3.3. LabsLand radioactivity remote experiment.
Source: www.labsland.com.

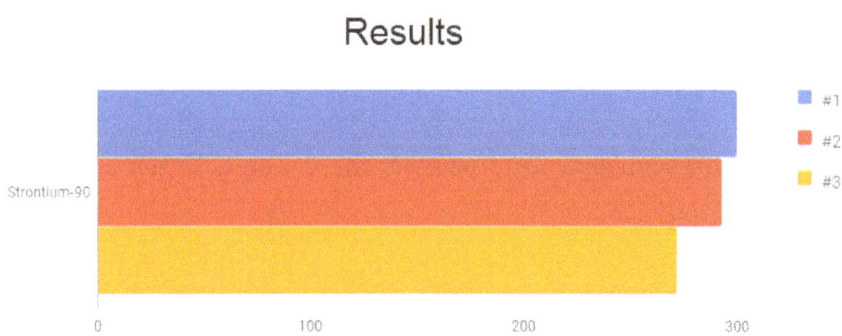

Figure 3.4. Results of the radioactivity remote experiment.
Source: www.labsland.com.

The interface provides, as well as the graph, a file with the angular position of the pendulum for each instant.

In this experiment, the student can only control the start position of the pendulum (the angle), but in other experiments the student can add a can of coke at different parts of the pendulum.

Radioactivity

The student chooses a radioactive material in order to count the number of radioactive particles detected. The student can choose the duration of

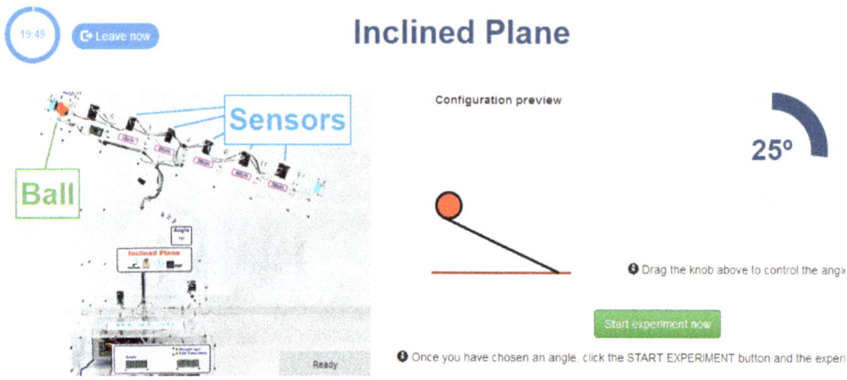

Figure 3.5. LabsLand inclined plane remote experiment.

Source: www.labsland.com.

Sensors	1st	2nd	3rd	4th	5th	6th
d (cm)	6	16	26	36	46	56
t (ms)	161	299	398	479	550	612

Figure 3.6. Results of the inclined plane remote experiment.

Source: www.labsland.com.

the experiment, its distance and whether there is an absorber and what it is (see Fig. 3.3).

The interface can clearly be improved upon and the student only receives visual and graphic feedback, having to note down the number of impacts (see Fig. 3.4). It is noteworthy that this experiment is in Australia.

The student can control all the experiment variables: material, duration, distance and absorber.

Inclined plane (kinematics)

The student selects the angle of inclination of the plane before launching the ball and to measure its behaviour in time (see Fig. 3.5).

The graphic interface is correct and returns to the student the time or moment for each point of the plane (see Fig. 3.6). The student must copy the experiment data to subsequently calculate the ball's speed and acceleration in its fall in order to test Newton's laws and the value of gravity, g.

The student should be able to control some other experiment variables, not only the angle. For example: plane length, ball mass and plane friction. These experiments are part of the work in progress.

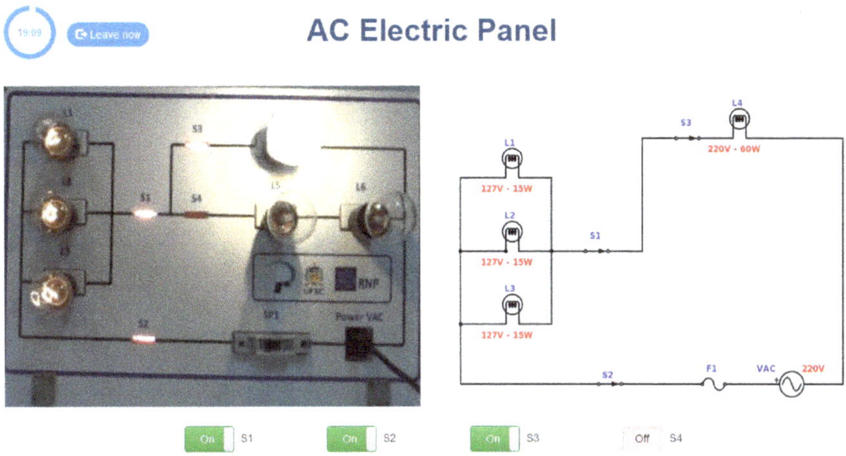

Figure 3.7. LabsLand alternating current circuits remote experiment.

Source: www.labsland.com.

AC circuits

The student has a panel with bulbs and switches in order to see how the current affects the closed switches (see Fig. 3.7).

The interface and the student's control of variables are correct. The student only receives visual feedback, and nothing else is necessary. This is a difficult experiment to understand as the current that passes through each bulb depends on the power, temperature and resistance of each bulb according to a nonlinear behaviour in a zero-sum model.

Planarians

The student can observe a planarian's movements in different kinds of water — pure, sugared, with caffeine, with taurine and with ginseng — in order to observe how the different stimulants affect its behaviour (see Fig. 3.8).

The interface and the student's control of variables are correct. The student uses the counter to measure the planarian's activity.

VISIR: Virtual Instruments Systems in Reality

The student can assemble electronic circuits with resistors, capacitors, etc. to power them and measure their behaviour with a multimeter and an oscilloscope (see Fig. 3.9).

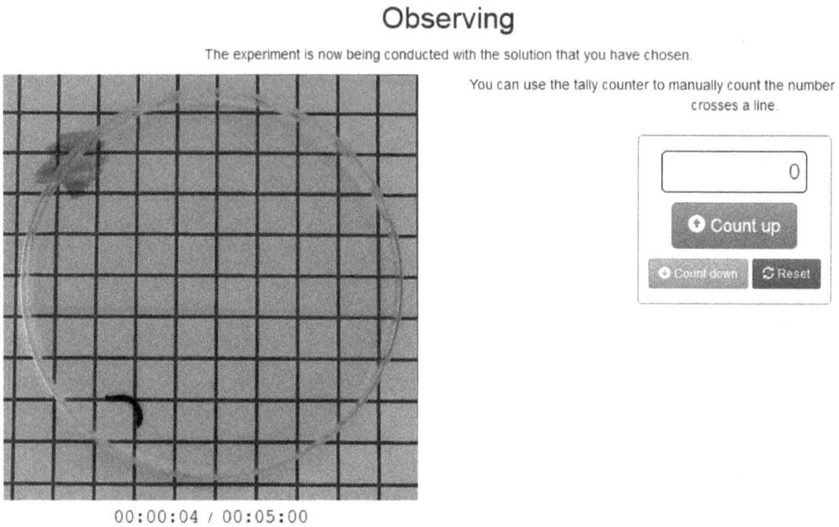

Figure 3.8. LabsLand planarian remote experiment.
Source: www.labsland.com.

VISIR is perhaps the most powerful, spectacular and frequently used remote experiment in the world, and it has won various awards. It costs over 40,000 euros. VISIR is deployed in Sweden (two examples), Spain (two examples), Portugal (two examples), Austria, Argentina (two examples), Brazil (three examples), the USA, Germany, Georgia and Costa Rica.

VISIR multiplexes its signals, and consequently a single model can offer a service to various students at a time (it is multiuser, though not deferred), up to a maximum of 50–60 students. In other words, the remote experiment is batch, although for the user it is interactive.

VISIR does not allow the collaboration among students in the same circuit: one circuit for one student. In this sense, NetLab (Nedic *et al.*, 2003) allows several students to work with the same circuit: NetLab allows collaborative work.

The student controls all the input variables and observes the output variables, and the interface also accurately simulates a workstation in a hands-on laboratory.

Arduino Robot and Arduino Board

In both cases, the student has to write a programme that implements an algorithm which controls a robot or another devices like LEDs, motors, LCDs, etc. (see Fig. 3.10).

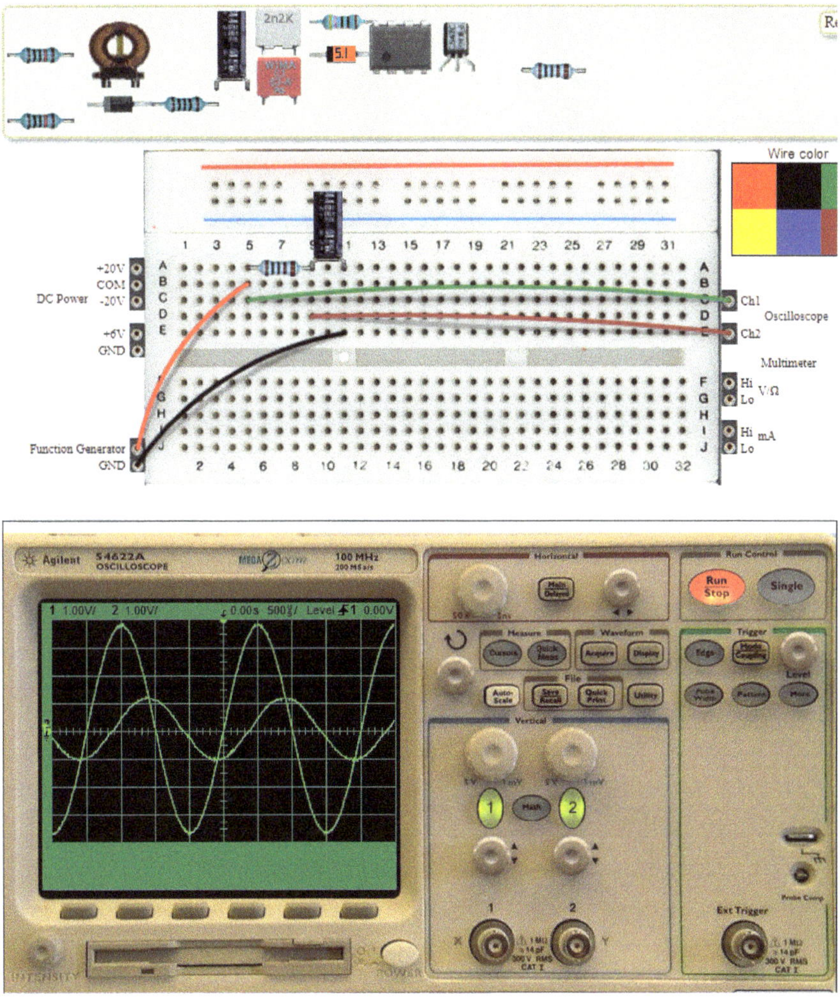

Figure 3.9. LabsLand VISIR remote laboratory.

Source: www.labsland.com.

The previous experiments were 'scientific' — discovering and explaining nature's behaviour in different spheres — but this is technological and its objective is to control a robot.

Of most interest is the fact that students can write their algorithms in different languages (C and Blockly) and that compilation and execution are performed online, so the student does not need to install any additional software.

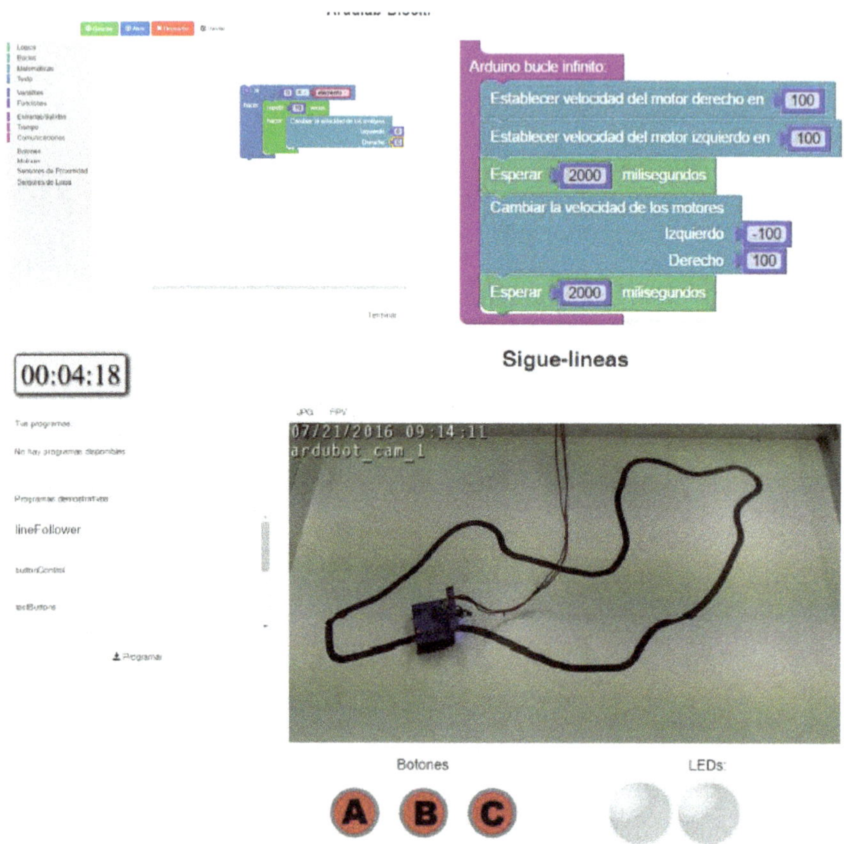

Figure 3.10. LabsLand programmable robot remote experiment.
Source: www.labsland.com.

A robot can only be controlled by one student at a time, and therefore is not nor can it be multiuser. This situation moves in two directions. On the one hand, the experiment manages a queue and assigns three minutes to each student, and on the other there are nine robots available (scalability) to offer a service to various students at a time.

At the time of writing, there are multiple robots in LabsLand in Spain, the University of Deusto in Spain, the University of Fort Hare in South Africa, UNAD in Colombia, and soon in UNED in Costa Rica. Students from UNED in Costa Rica — or from a school in Costa Rica or from a university affiliated to UNED — write a code and every time they try to test it in a real robot, LabsLand first tries to assign them to the robots in

Costa Rica. If they are already in use or unavailable (due to maintenance or similar), automatically they go to any of the available robots elsewhere in the LabsLand network. If all the robots in the LabsLand network are already in use, the student will be queued until one becomes available. Given that sessions are, by design, very short, long queues are hardly ever formed. This federation model is technically complex but very reliable and scalable.

In the case of the Arduino, the experiment is very similar: the student writes code and uploads it to the Arduino, manipulates the inputs and observes their effect on the outputs connected to the board (see Fig. 3.11).

In this remote laboratory, the instructor's role is crucial, but no less important are the students, as their creativity and ingenuity when programming is the most relevant aspect. Technically this is a very complex laboratory, which is explained in a very short time and offers considerable scope in the classroom, not only for a week, as in the case of the pendulum or other similar experiments, but throughout an entire subject.

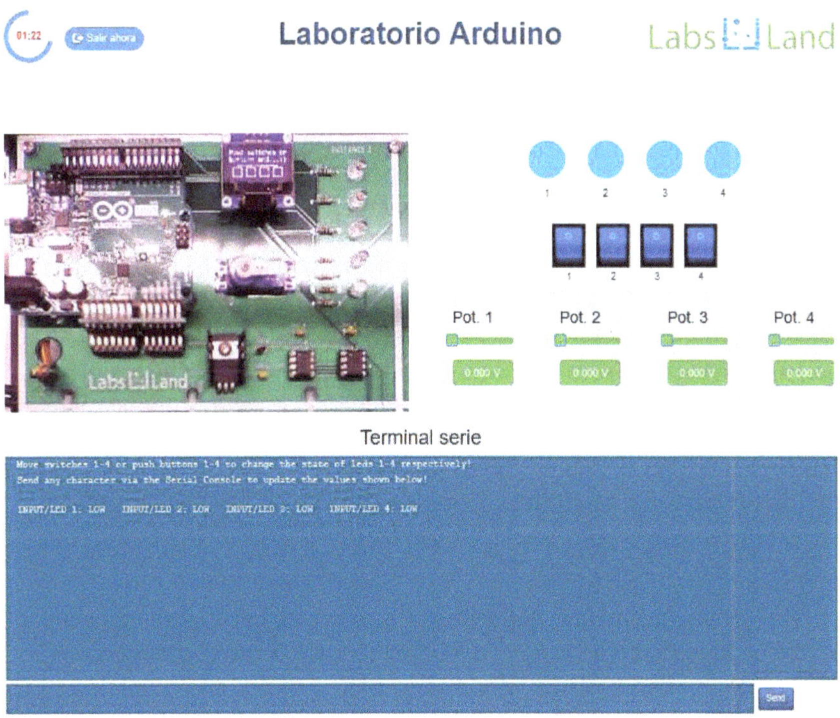

Figure 3.11. LabsLand programmable Arduino remote experiment.

Source: www.labsland.com.

3.2.2 *Analysis of LabsLand experiments*

Having seen and examined the aforementioned experiments in detail, it makes sense to describe them using a subset of the criteria from Sections 2.1 and 2.3.

The criteria used are:

- **UN (Universality):** The experiment is accessible in any scenario.
- **Q/B (Queue/Booking):** If the experiment is occupied, the student joins a queue or makes a booking.
- **MU (Multiuser):** The experiment can be accessed by various students at a time.
- **TD (Time Deferred):** The student accesses a recorded experience.
- **O/B (Online/Batch):** The student interacts and receives feedback in real time or after a certain time (milliseconds, seconds, hours).
- **Pa/Pr/N (Parametrizable or Programmable or Nothing):** The experiment can be parametrised or programmed.
- **C/S (Combinational or Sequential).**
- **C&O (Controllability and Observability):** The student observes and controls all the relevant experiment variables.
- **F&D (Feedback and Data):** The student benefits from quality feedback from the experiment, in the form of both images and data.
- **LA (Learning Analytics):** The experiment records the student's activity.

If certain features do not appear it is because LabsLand experiments do not include them, and they are related to virtual and augmented reality.

Table 3.2 some of the aforementioned criteria are subjective, but nonetheless, the following table lists and classifies the remote experiments per the above-mentioned categories.

Table 3.2. Analysis and classification of remote experiments.

	UN	Q/B	MU	TD	O/B	C/S	P/P/N	C&O*	F&D	LA
VISIR	X	—	X	—	B	S	Pr	X	x	X
RADIO	X	—	X	X	B	C	Pa	X	x	X
PENDULUM	X	—	X	X	B	C	Pa	X	X	X
KINETICS	X	—	X	X	B	C	Pa	X	X	X
AC CIRCUITS	X	—	X	X	O	C	Pa	X	X	X
ARCHI	X	—	X	X	B	C	N	X	x	X
PLANAR	X	—	X	X	B	C	N	X	X	X
ROBOT&ARDU	X	Q	—	—	O	S	Pr	X	X	X

Note: *An X means that the remote experiment is completely C&O, while an x means that it is partially C&O.

3.2.3 *ISES experiment*[2]

ISES, coordinated by F. Lustig in Czech Republic for over 10 years, offers approximately 20 physics experiments, all of a scientific nature, where the student parametrizes the experiment, launches it and observes it in order to draw conclusions (Lustig *et al.*, 2018).

The experiments are parametrizable, in real time, single-user and managed by a queue. They are all simple and controllable, and the results can be stored.

One example is electromagnetic induction, where the student can control supply voltage and the number of turns in the coil (see Fig. 3.12). The experiment returns data via graphics and Excel files.

Another experiment is the rectifier. The features are similar to the aforementioned experiments, but this case is comparable to the VISIR. Figure 3.13 shows the ISES interface, and Fig. 3.14 shows how it would be in VISIR.

In ISES, the student clicks on the interface options (see Fig. 3.13), but does not assemble or connect the circuit or the instrumentation, while in VISIR (see Fig. 3.14) the student is responsible for everything. Which experiment or approach is better?

[2] https://www.ises.info/index.php/en/laboratory.

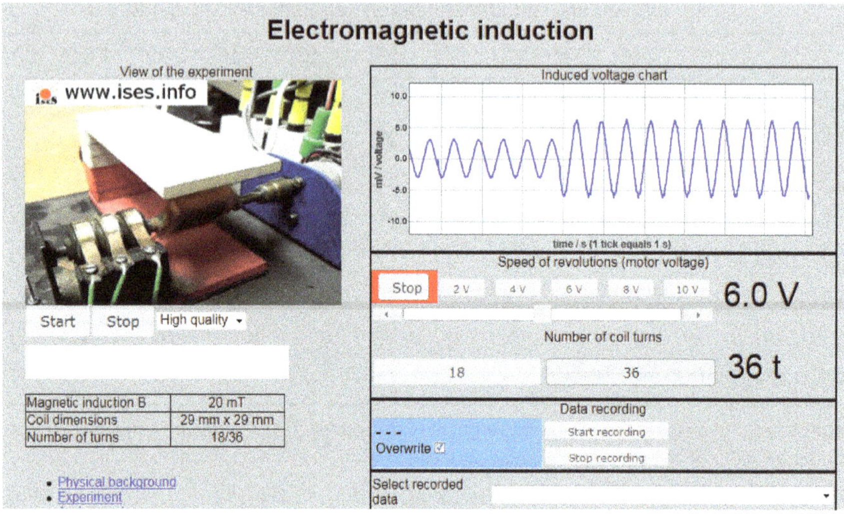

Figure 3.12. ISES electromagnetic induction remote experiment.
Source: https://www.ises.info/index.php/en/laboratory.

Other ISES experiments make it possible to analyse the photoelectric effect and the Heisenberg uncertainty principle, but are not always available or responsive (see Fig. 3.15).

3.2.4 *UNED's UNILabs*[3]

UNED's UNILabs, coordinated by Sebastián Dormido, offers 24 remote experiments integrated into Moodle and documented, most oriented towards systems control and at the university level. They are complex and expensive experiments and are the state-of-the-art in remote experimentation in control systems (Heradio *et al.*, 2016b).

They are online, single-user experiments, operated via booking and generally controllable by means of a PID controller or similar. Many of the experiments feature augmented reality (Galan *et al.*, 2019). UNILabs also provides a simulator based on EjsS — Easy Java/JavaScript Simulations — (Esquembre, 2004) for each experiment so the student can practise and learn offline.

[3] https://unilabs.dia.uned.es/.

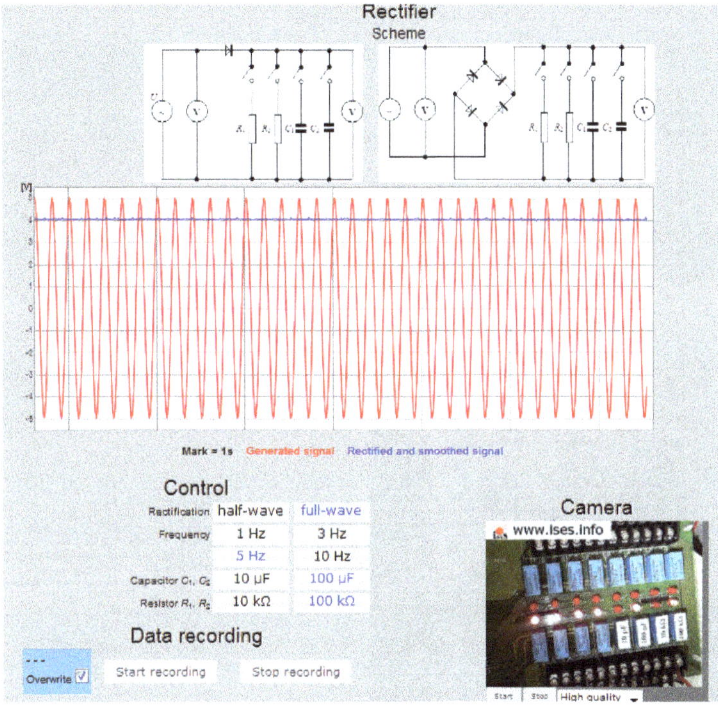

Figure 3.13. ISES rectifier circuit remote experiment.

Source: https://www.ises.info/index.php/en/laboratory.

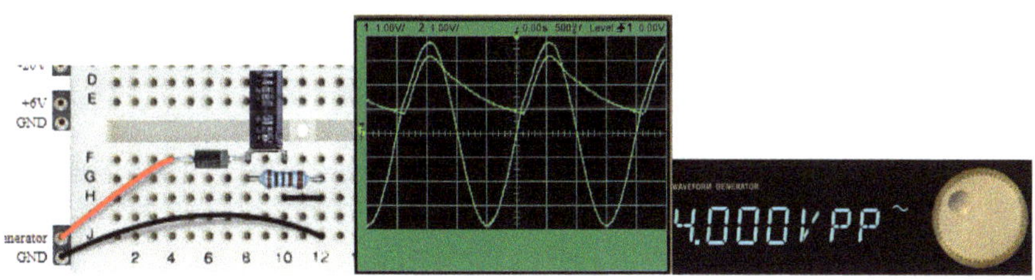

Figure 3.14. Rectification remote experiment with VISIR.

Source: www.labsland.com.

Figure 3.16 shows the interface that controls the level of the three tanks. The student parametrizes a controller and observes the latter's behaviour in a graph and via augmented reality.

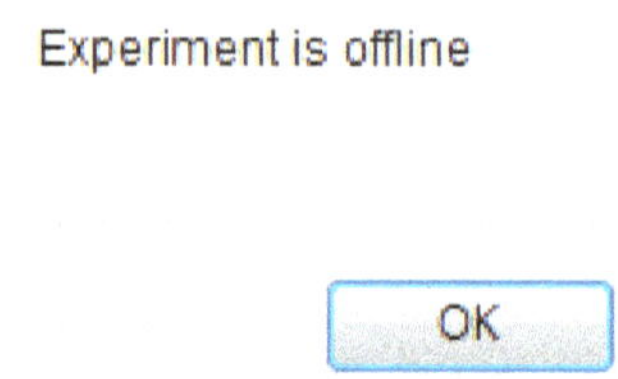

Figure 3.15. Message from ISES saying that the selected experiment is offline.
Source: https://www.ises.info/index.php/en/laboratory.

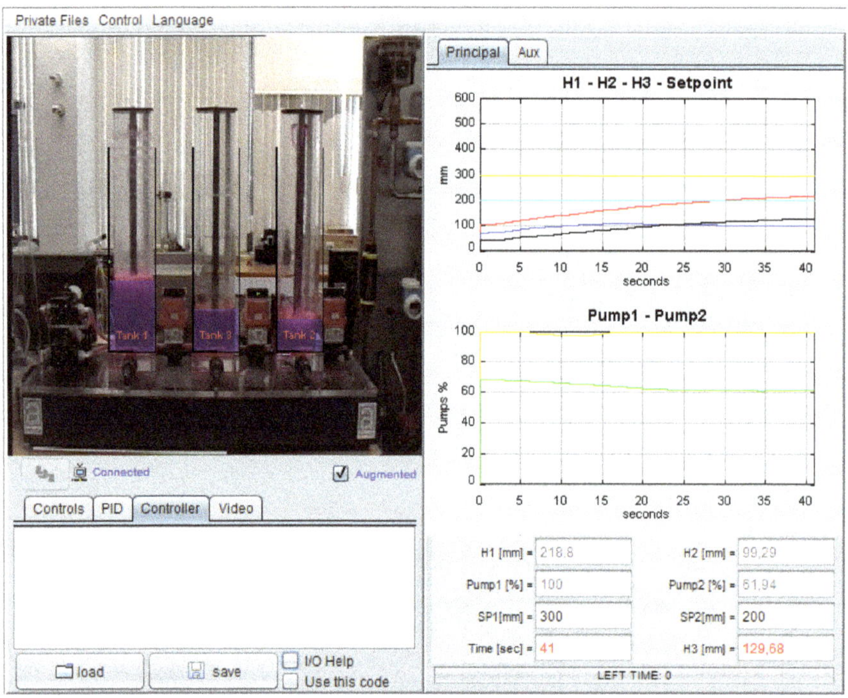

Figure 3.16. UNILabs process control remote experiment.
Source: https://unilabs.dia.uned.es/.

Another UNILabs experiment enables the student to learn about the reflection and the refraction of light by means of a laser beam and augmented reality. The student chooses the angle of incidence and can see on the interface the angles of reflection and refraction highlighted with augmented reality (see Fig. 3.17).

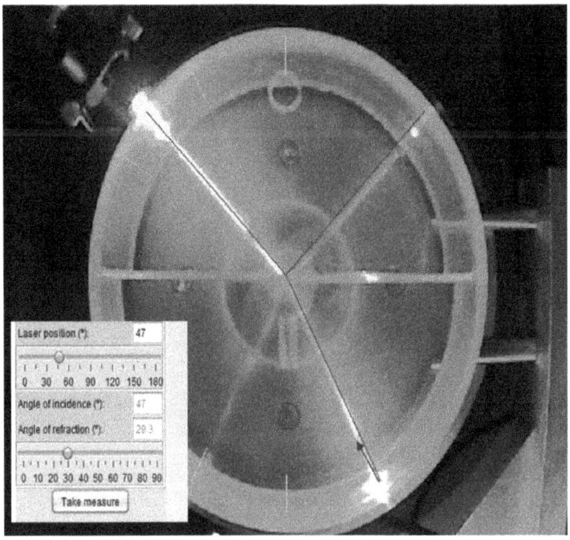

Figure 3.17. UNILabs diffraction remote experiment.
Source: https://unilabs.dia.uned.es/.

3.2.5 *Experiments at the RCL of the University of Munich*[4]

For 15 years, RCL (Gröber *et al.*, 2007) has been offering a wide range of experiments focused on university or high-school-level basic physics. These are very similar to the ISES experiments.

RCL offers approximately 20 physics experiments, all of a scientific nature, where the student parametrises, launches and observes the experiment in order to draw conclusions. The experiments do not return data, the experiment can only be observed via the webcam.

The experiments are parametrizable, in real time and for a single user. The main problem is that they do not manage queues, so if a student wishes to participate in an experiment that is already occupied, not only can they not do so, but moreover they do not know what is happening.

The photoelectric effect experiment enables the student to parametrize two variables to see the voltage value reached on the multimeter (see Fig. 3.18).

[4]http://rcl-munich.informatik.unibw-muenchen.de/.

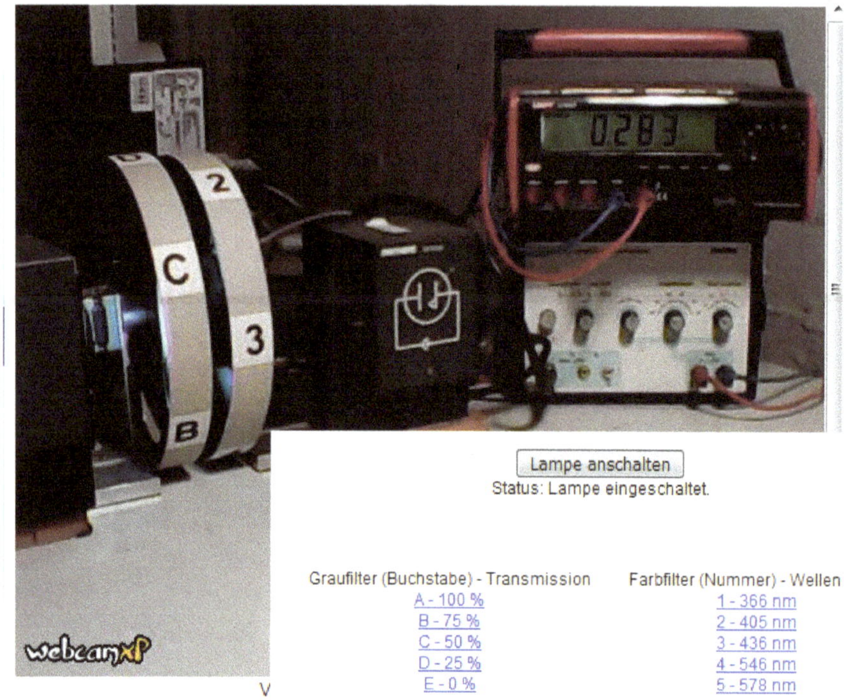

Figure 3.18. RCL photoelectric effect remote experiment.
Source: http://rcl-munich.informatik.unibw-muenchen.de/.

Figure 3.19 shows an experiment for experimentally calculating the speed of light. Unfortunately, the interface is neither very user-friendly nor intuitive, but the technical support is excellent.

3.2.6 *Experiments at the OpenEngineering Laboratory Open University*[5]

A field with an unusually low number of remote experiments is that of natural sciences. The Open University (UK) was funded by the British Government and it has developed a set of remote experiments that can be found on its website (https://learn5.open.ac.uk/course/view.php?id=2).

[5]https://learn5.open.ac.uk/course/view.php?id=2.

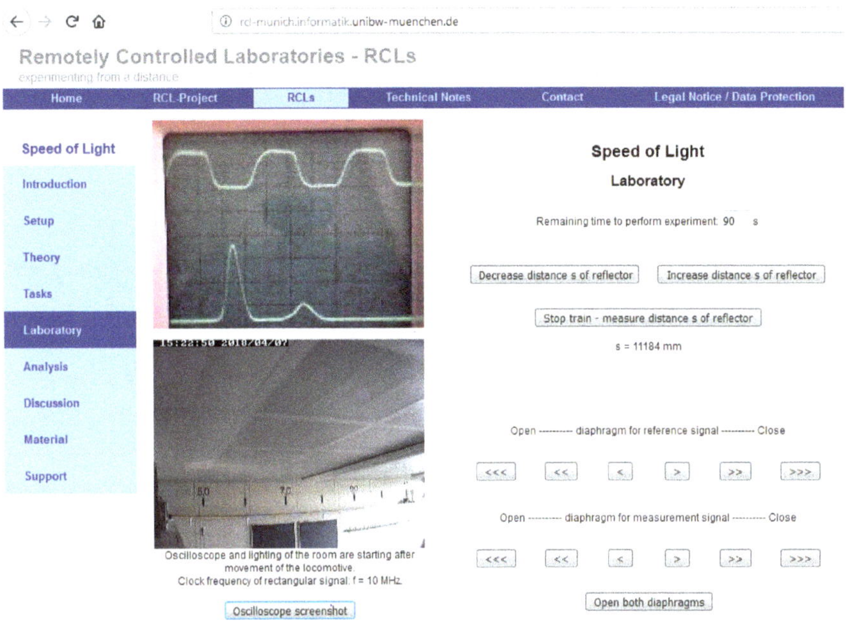

Figure 3.19. RCL speed of light remote experiment.

Source: http://rcl-munich.informatik.unibw-muenchen.de/.

The OU offers over 50 scientific experiences: remote, virtual, interviews, etc. Most of the remote experiences are not available or require users to be enrolled on an OU course in order to access them via a booking system. Each remote experiment is designed by a different team, so their features vary considerably (see Fig. 3.20).

The OU offers various experiments using microscopes. These experiments fall within the Data Set category, as the student accesses, observes and analyses in real time an image created beforehand. The experience is real since students access a digital microscope that they can manipulate, where the only thing they cannot do is remove the sample from the box and place it under the microscope. The OU offers many kinds of geological and biological samples. Figure 3.21 shows an OU microscope which when accessed shows a message indicating that the experiment requires Flash, which currently is not recommended.

Another OU experiment allows the student to analyse the presence of nitrates in different water samples using a spectrometer. This experiment

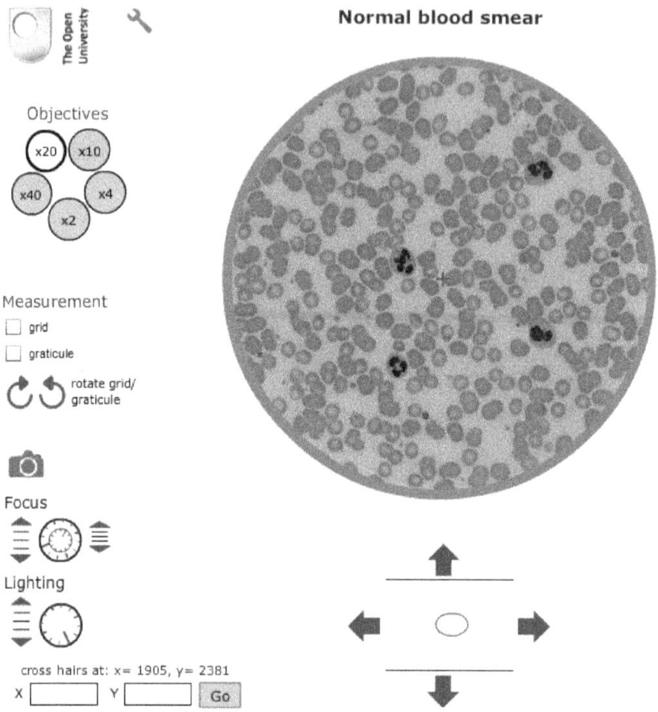

Figure 3.20. Open University microscope remote experiment to observe a blood sample. *Source*: https://learn5.open.ac.uk/course/view.php?id=2.

is Data Set with an interface that replaces the job function of a laboratory technician (see Fig. 3.22).

In another experiment, the OU enables the student to experiment with electrons controlling the equipment's voltage and current (see Fig. 3.23). The interface and the user experience give the impression that the experiment is online, but the fact that it is not necessary to book it suggests that it is a virtual laboratory with an excellent interface. This is a good example of how difficult it can be to differentiate between one type of experiment and another, although really we should consider whether or not it produces learning results.

The OU places emphasis on guaranteeing that the user's experience will be positive and, to this end, invites users to test the equipment before accessing an experiment (see Fig. 3.24). This situation is in line with the concept of trustability seen earlier.

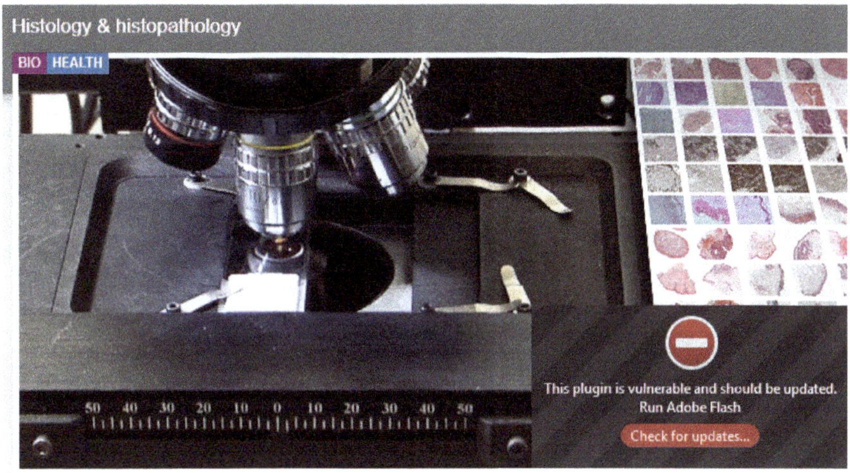

Figure 3.21. Problems with Adobe Flash in the Open University microscope.
Source: https://learn5.open.ac.uk/course/view.php?id=2.

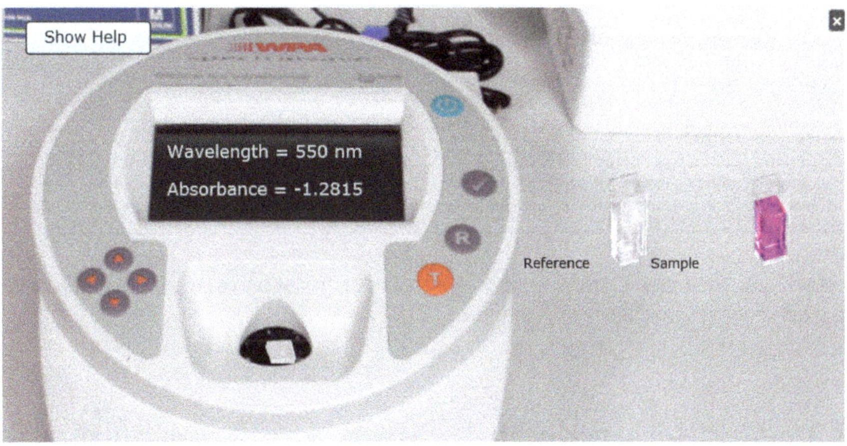

Figure 3.22. Open University water analysis remote experiment.
Source: https://learn5.open.ac.uk/course/view.php?id=2.

To summarise, the OU experiments are remote, Data Set and virtual and are controlled via a booking system (the remote experiments) They have sophisticated interfaces, problems of universality in some cases, are read-type and do not permit programming or advanced control, and in general do not offer additional data via excel files and graphics.

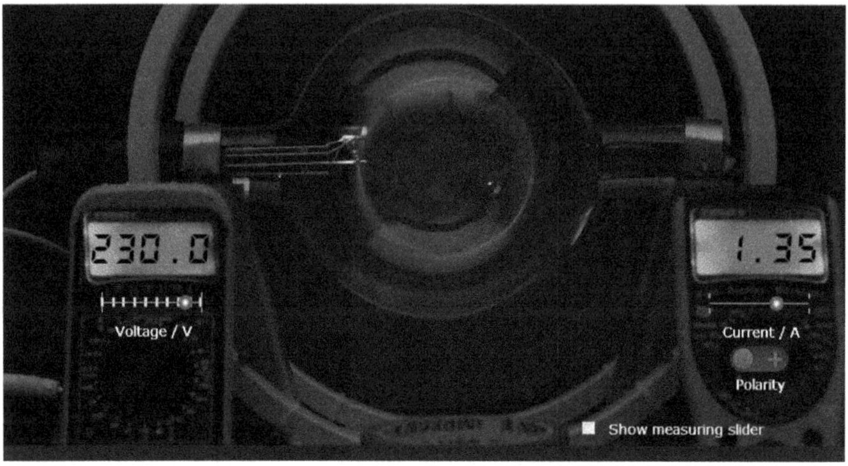

Figure 3.23. Open University electrons remote experiment.

Source: https://learn5.open.ac.uk/course/view.php?id=2.

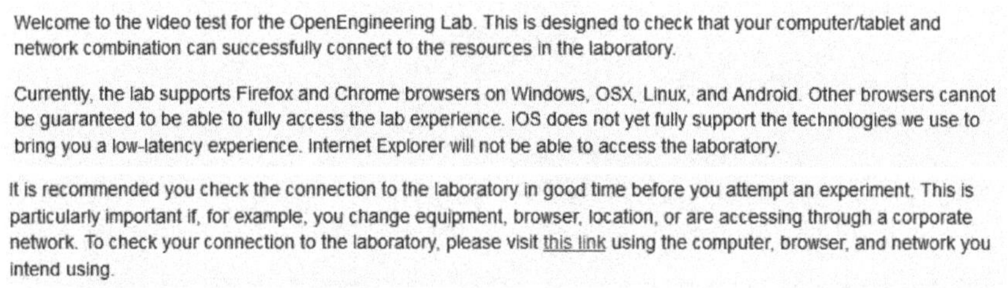

Welcome to the video test for the OpenEngineering Lab. This is designed to check that your computer/tablet and network combination can successfully connect to the resources in the laboratory.

Currently, the lab supports Firefox and Chrome browsers on Windows, OSX, Linux, and Android. Other browsers cannot be guaranteed to be able to fully access the lab experience. iOS does not yet fully support the technologies we use to bring you a low-latency experience. Internet Explorer will not be able to access the laboratory.

It is recommended you check the connection to the laboratory in good time before you attempt an experiment. This is particularly important if, for example, you change equipment, browser, location, or are accessing through a corporate network. To check your connection to the laboratory, please visit this link using the computer, browser, and network you intend using.

Figure 3.24. The Open University laboratory connection test proves the OU's dedication to guaranteeing a positive user experience.

Source: https://learn5.open.ac.uk/course/view.php?id=2.

3.2.7 *RexLab experiments*[6]

RexLab, (Silva *et al.*, 2016), offers 12 remote experiments with varied subject matter, but focus in particular on electricity and computing. It is currently developing and testing 10 more experiments. The research team has acknowledged expertise, and RexLab has been active since 2008 and is one of the international references in remote experimentation.

[6]https://rexlab.ufsc.br/en/.

Figure 3.25. RexLab microscope remote experiment.
Source: https://rexlab.ufsc.br/en/.

The laboratory is deployed in Brazil, which makes quality access difficult, compromising the trustability of experiments and their use in the classroom. This situation led to LabsLand reaching an agreement with RexLab to convert some of them into Time-Deferred or Pre-recorded experiments, in order to take full advantage of their power, maximising their accessibility.

LabsLand's kinematics (inclined plane), AC circuits and Microscope experiments (see Fig. 3.25) belong to RexLab and provide a good description of its characteristics.

The RexLab experiments are remote, are controlled via queues and are single user. Their interfaces are correct and do not present problems of universality. They offer data in files and are not very stable, owing to connectivity problems resulting from poor network quality.

3.2.8 *Online Experimenta of FEUP, OE@FEUP*[7]

The FEUP remote laboratory in Porto (Portugal) (Restivo *et al.*, 2011) is renowned for both its results and the activity of the research group led by Teresa Restivo.

[7]https://remotelab.fe.up.pt/.

This laboratory is articulated around three axes: remote experiments, virtual reality experiments and augmented reality experiments. Thus, if a remote experiment is not available, the student can use the virtual version. In many cases, before accessing the remote experiment, the student should have tried the virtual one.

FEUP offers 12 remote experiments related to mechanics, materials technology and medical and civil engineering. Of these 12, four use LabVIEW and eight are HTML. The LabVIEW version is number 7, and therefore requires a plugin, which *a priori* renders it totally inadvisable in an educational centre. OE@FEUP also reports problems with Internet Explorer.

The other eight remote experiments are already accessible via HTML without any need for a plugin. They are all connected to Moodle and are therefore accompanied by didactic material.

In lamp temperature control, the student can see how an ON/OFF control evolves based on a lamp and a fan. The image shows the temporal evolution of the temperature. The graph shows the effect of the ON/OFF control (see Fig. 3.26).

In this experiment, the student cannot write the controller program, and only observes its effect. This approach is very different from that of other experiments seen earlier, which were either controllable or at least

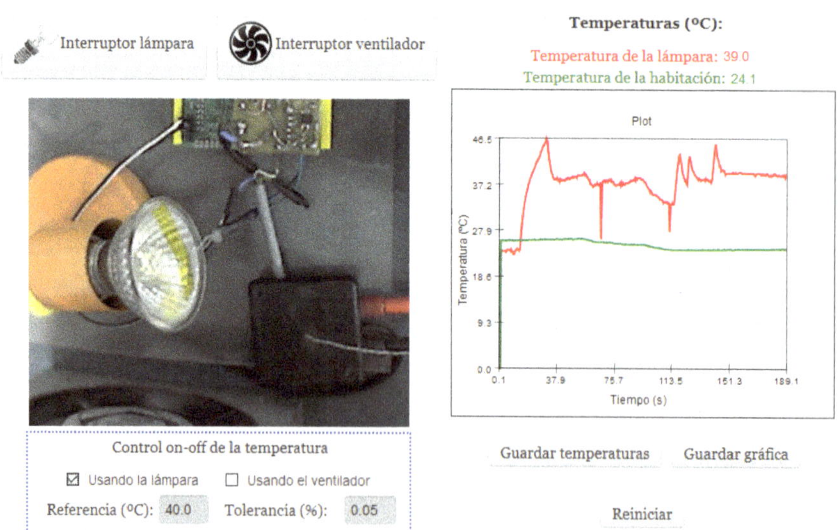

Figure 3.26. OE@FEUP ON/OFF control remote experiment.
Source: https://remotelab.fe.up.pt/.

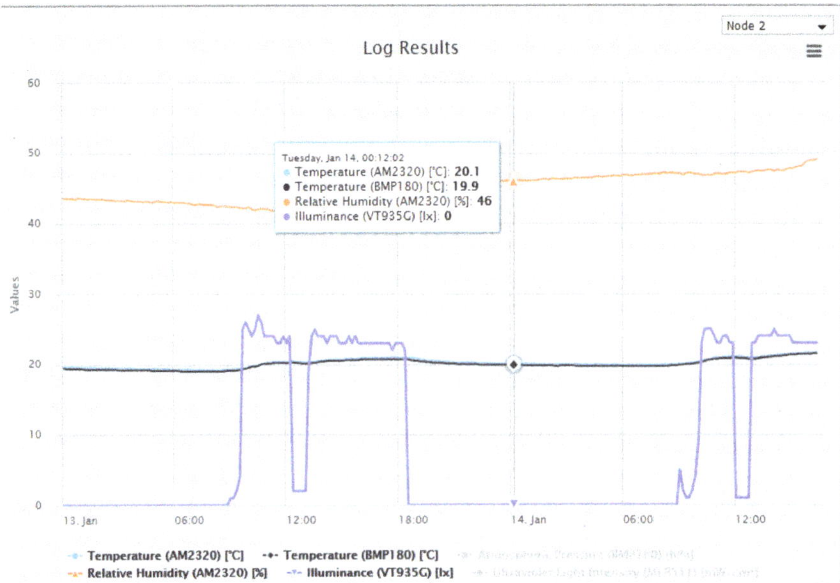

Figure 3.27. OE@FEUP comfort measures remote experiment.

Source: https://remotelab.fe.up.pt/.

parametrizable in their values, and not only in ON/OFF as in the OE@ FEUP experiment.

Within the sensored category, OE@FEUP provides access to the data recorded by temperature, humidity and light sensors in a laboratory. In this case, the student neither interacts nor requests anything from the remote experiment, but simply observes and analyses the data (see Fig. 3.27).

3.2.9 *Lisbon IST e-lab experiments*[8]

The Technical Institute of Lisbon University is home to the e-lab remote laboratory which, coordinated by Horacio Fernandes, (Fernandes *et al.*, 2013), contains an excellent set of basic and advanced physics remote experiments (see Fig. 3.28).

Figure 3.29 shows the pendulum that the student can configure and launch in Faro (Portugal) as well as various images: evolution of the

---

[8] http://www.elab.ist.utl.pt/.

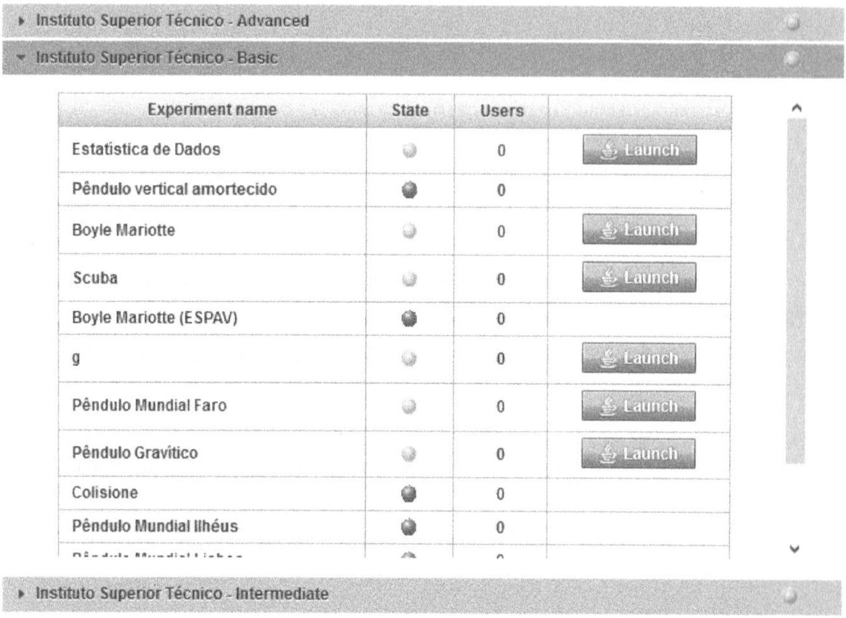

Figure 3.28. State of the remote experiments at the UTL e-lab laboratory.
Source: http://www.elab.ist.utl.pt/.

pendulum and histogram to calculate the pendulum period (see Fig. 3.29). All the experiments include very detailed educational material.

e-lab offers 25 remote experiments, of which at the time of access only half were available. If their main advantage is the high quality of the experiments, the enormous disadvantage is that they require the use of Java, the opening of several ports and removal of firewalls. This is unlikely to be permitted by an educational centre's IT service. If it was complicated to use the LabVIEW v7 plugin required by OE@FEUP, what e-lab requires is almost impossible.

3.2.10 *LaboRem experiments*[9]

The University of Pau, with Franck Luthon and Benoit Larroque, offers a very special experiment at its Bayonne campus in France. The subject

[9] https://www.iutbayonne.univ-pau.fr/tice/2015/04/19/laborem-laboratoire-a-distance-a-liut-de-bayonne/.

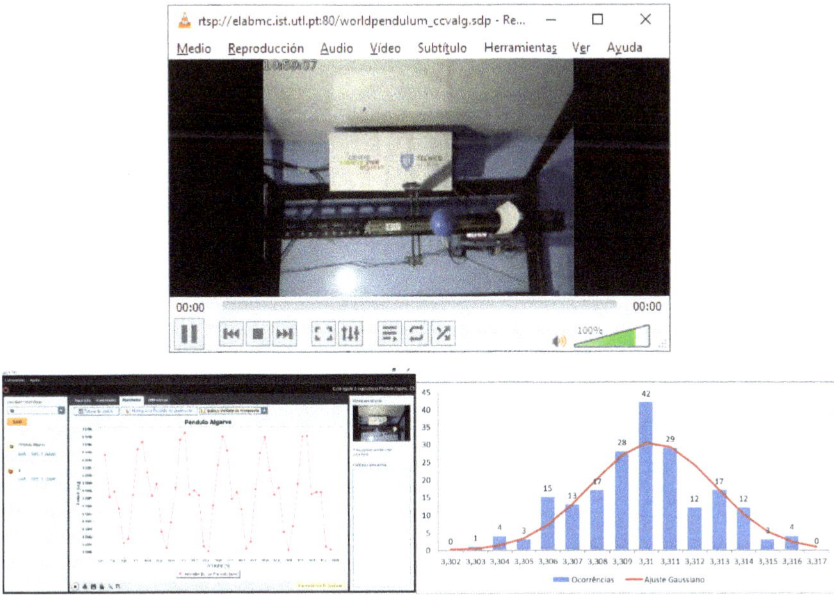

Figure 3.29. UTL e-lab pendulum remote experiment and results.
Source: http://www.elab.ist.utl.pt/.

matter is typical, as it offers students the possibility of experimenting with 16 electric and electronic circuits situated on a breadboard and partially configurable. For example, on an RC circuit it is possible to change the capacitor value.

What is amazing about this remote experiment is that it is configured by means of a robot that takes a capacitor from a set thereof and places it in an area connected by relays to the RC circuit. In other words, the student sees the manipulation of the circuit and not only its effect, as in other remote experiments. This interface certainly favours student immersion (see Section 4.1).

Figure 3.30 shows the appearance of the remote experiment, but it is better seen on video https://www.youtube.com/watch?v=2bsfeJIL5GU.

Apart from the aforementioned feature, LaboRem is noteworthy because it uses PyScada (developed by the Technical University of Berlin) to control the measurement and power equipment. PyScada is open source and permits control of equipment without needing to use LabVIEW or other proprietary programs and protocols.

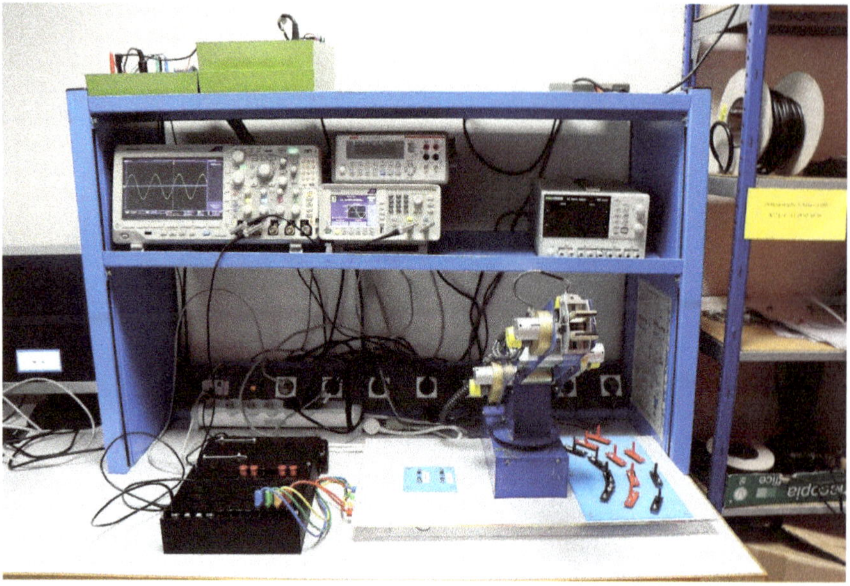

Figure 3.30. LaboRem circuits with robot remote experiment.

Source: https://www.iutbayonne.univ-pau.fr/tice/2015/04/19/laborem-laboratoire-a-distance-a-liut-de-bayonne/.

3.2.11 *GOLDi: The Grid of Online Laboratory Devices Ilmenau*[10]

GOLDi is a project of the Technical University of Ilmenau in Germany, coordinated by Karsten Henke and Heinz-Dietrich Wuttke. GOLDi is a hybrid online lab devoted to programmable devices (Henke *et al.*, 2016) (Henke *et al.*, 2019).

The remote lab offers a set of programmable devices that can control a set of electromechanical models (https://www.goldi-labs.net/index.php?Site=114) (see Fig. 3.31). The user writes the program in C, assembler, etc. and uploads it in a microcontroller, FPGA or PLC, then connects this programmable device with a physical model like an elevator to see if the program runs well or needs to be improved. GOLDi is a hybrid remote lab, which means that the user can control a real physical model or can control a virtual/simulated model to test whether the control algorithm is correct or not.

[10] https://www.goldi-labs.net/.

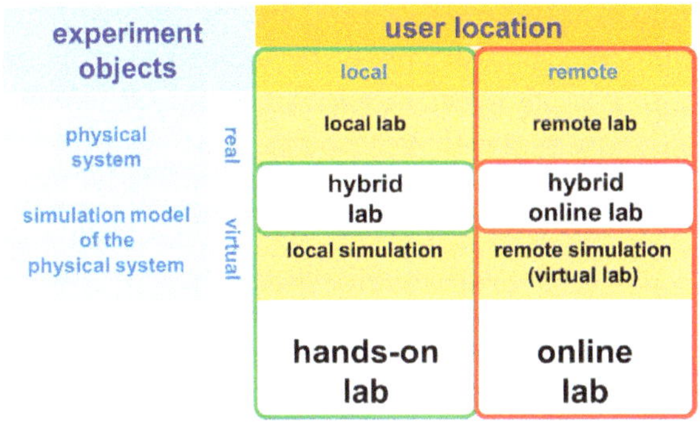

Figure 3.31. GOLDi remote lab description.
Source: https://www.goldi-labs.net/.

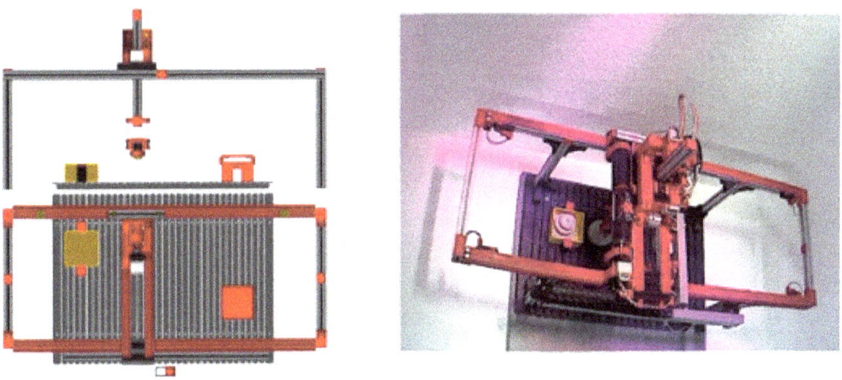

Figure 3.32. Virtual and physical model of GOLDi.
Source: https://www.goldi-labs.net/.

GOLDi can currently be regarded as state-of-the-art in remote experimentation with programmable devices, but requires some improvements, especially with the use of https protocols (see Fig. 3.32).

3.2.12 *Physics Lab Farm*[11]

The ISEP Laboris research team in Oporto (Portugal), led by Gustavo Alves (http://www.cieti.isep.ipp.pt/index.php?page=laboris), offers various

[11] http://physicslabfarm.isep.ipp.pt.

remote experiments in the Physics Lab Farm, among which VISIR stands out. Apart from the technical effort, the Laboris team is notable for its holistic vision of remote experimentation, with particular emphasis on its pedagogical and organisational aspects. This vision has enabled it to successfully coordinate two European projects, VISIR+ (Alves *et al.*, 2016) and RextNet, and participate in many others. There is no doubt that they are a reference in remote experimentation.

Among their experiments, the one presented in Fig. 3.33 is notable for its electromechanical complexity. In this case, the student can launch a steel ball in order to observe its evolution in time (Alves *et al.*, 2013). Students participate in the entire process, as they raise the ball to a chosen height, select the launch angle and finally launch the ball. A system of sensors measures the ball's position at every moment. The results obtained enable students to study the launch.

Figure 3.33. Laboris-ISEP projectile launching remote experiment.

3.2.13 *Trustability and reliability of remote experiments*

In previous descriptions, we have seen that not all remote experiments are available. How many laboratories are available in general? What level of trustability and reliability exists?

We cannot consider the entire body of remote experiments because this is unknown and is not indexed, but we can use the GoLab ecosystem as a reference. The www.golabz.eu webpage offers the user remote and virtual laboratories. In total, over 500 online experiments are offered, at the moment of access (July 2019), 63 of which are remote experiments, provided mainly by universities and research teams: RexLab in Brazil, CUAS in Austria, UNED and UDeusto in Spain, University of Munich in Germany, etc. Of the 63 experiments, 50% are operational for users (32 experiments), 15% (10 experiments) are accessible but not operational, and as many as 30% (21 experiments) are not even available. With regard to providers of experiments, the RCL of the University of Munich provides 12 experiments, eight of which are available, and four are not; RexLab of UFSC of Brazil offers 10, eight of which are operational; and WebLab-Deusto offers four, all of them operational. This is just one example of the technical complexity of a remote laboratory that shows how difficult it is to remain sustainable.

In the same vein, the ISES (www.ises.info) is the only one offering specific hardware service for the design of remote laboratories, and in turn offers a set of 18 remote experiments, 12 of which are available (66%), and six are not (33%), although one is advised as to consider whether they are offline or under maintenance.

This type of random analysis of different remote laboratories should give us some idea of their complexity, but this should in no way affect their usefulness or suitability in the classroom. This analysis should clarify and emphasise the fact that providing a quality remote experimentation service in the classroom is a complex question and requires professional actors.

Chapter 4

Pedagogical and Educational Value of Remote Experimentation

This chapter has three sections. The first discusses the suitability of remote experimentation in order to achieve learning results, to which end there is presentation and discussion of the results obtained by different researchers with the ultimate objective of enabling the interested reader to replicate or adapt these studies in accordance with section three.

The second section describes the didactic approach of various accessible remote experiments. In the third section, the evaluation of remote experimentation is described.

4.1 Pedagogical Value of Remote Experimentation

A recurring issue in remote experimentation is knowing whether it produces learning results, whether it has a positive impact upon learning. This analysis is crucial for the implementation and use of remote experiments in educational centres.

In this situation, it is important to use the specific techniques of teaching evaluation. These, of course, tend to be in the hands of psychologists and educationalists, and not in the hands of the engineers who usually design the remote experiments analysed.

What follows is a brief presentation of the results obtained by different researchers.

The work published by Ma and Nickerson (2006) is a reference within the first 10 years of remote experimentation. This is a meta-study (survey) of the results obtained and published in 60 journal papers during the previous 10 years. Various elements of the publication can be highlighted: in the first place, the authors emphasise that 'It is hard to perform large-scale educational tests and hold factors such as instructor ability constant. It is also difficult to compare studies which focus on different scientific domains'. After the study, the authors conclude that 'There is a fair amount of evidence that simulated and remote labs are effective in teaching concepts'. The article highlights the fact that the researchers whose works have been reviewed mix up criteria and objectives, which makes it difficult to categorize findings. However, Ma and Nickerson establish various crucial elements in the analysis, two of which stand out: technology and the sensation of immersion.

From the technological point of view, as has been noted, laboratory equipment is already measured computationally, and even that computerised measurement reaches the Internet, so it is difficult to differentiate between hands-on laboratories and remote laboratories. That affirmation from 2006 is truer than ever today.

Ma and Nickerson then state that 'Second, the effectiveness of laboratories may be affected by how much students believe in them. Therefore, an understanding of presence, interaction, and belief may lead to better interfaces. Also, if belief proves important, then hybrid approaches might be contemplated, in which hands-on work is used at an early stage to build confidence in remote or simulated technology used in later teaching'. The sensation of immersion can be fundamental in order to take full advantage of the remote experiment, 'Furthermore, research in psychology suggests that the beliefs and experiences of students may be determined more by the nature of the interfaces than by the objective reality of the laboratory technology'. The article is very inspiring.

As mentioned earlier, it is worth returning to remote experiments that are recorded and deferred in time. When users access these experiments, they see a recording with which they interact in a reliable manner. This fact, recording and reproduction, can affect the user's 'feeling of immersion', and therefore, according to Ma, may affect the user's global experience.

Similar conclusions are established in Corter *et al.* (2007) and Nickerson *et al.* (2007). This makes up for the limited accessibility of this kind of experiment, which has already been noted by Lahoud and Krichen (2010).

Brinson (2015) embarked on a similar project to the one performed by the two aforementioned authors. As he explains, 'The purpose of this paper is to expand upon and update their findings to include literature published in and after 2005. More specifically, the focus herein is on comparative studies that offer direct, empirical measures of student achievement of learning outcomes of the newer NTL (experimental group) vs. TL (control group); very few such studies had been completed at the time of Ma and Nickerson's (2006) review'.

Brinson's meta-analysis emphasises that he is only going to analyse empirical experiments where the authors contrast the quality of traditional labs (hands-on) and non-traditional labs (virtual and remote labs). The study begins with a selection of almost 1300 articles, 56 of which he chooses according to different criteria. The end result is that 'This review presents the first attempt to synthesize recent (post-2005) empirical studies that focus on directly comparing learning outcome achievement using traditional lab (TL; hands-on) and non-traditional lab (NTL; virtual and remote) participants as experimental groups. Findings suggest that most studies reviewed (n = 49, 89%) demonstrate student learning outcome achievement is equal or higher in NTL versus TL'. Figure 4.1 summarises Brinson's analysis.

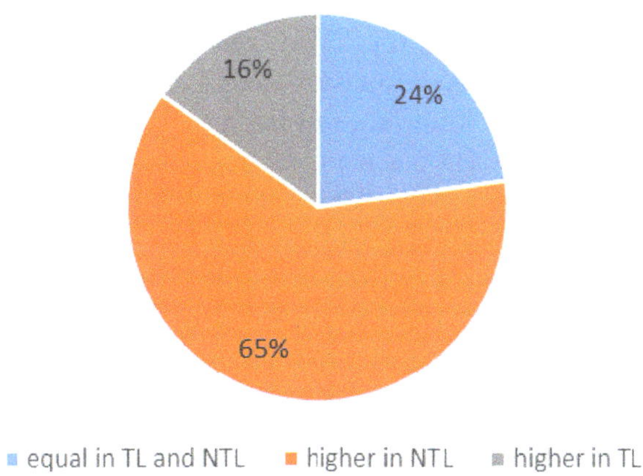

Figure 4.1. Learning outcome achievement according to Brinson (2015).

Beyond such a clear and categorical result, it is worthwhile noting a question that Brinson poses: Can an online course be accredited like a classroom course, engineering included? In his own words 'The need for research related to the effectiveness of non-traditional learning environments, including the NTL, is being recognized by both educators and policy makers. The United States Department of Education stated in a recent report that policy-makers and practitioners want to know about the effectiveness of Internet-based, interactive online learning approaches and need information about the conditions under which online learning is effective'. Brinson takes remote experimentation to a different scenario, where what is in doubt is not its didactic value but its 'administrative' value.

De Jong *et al.* (2013) published a short, hard-hitting work in the prestigious journal *Science*. The aim was similar to that of the earlier works: review what had been published by other researchers in order to analyse and compare physical (hands-on) and virtual laboratories. In this article, virtual laboratories include above all simulators, and to a lesser degree, remote laboratories. One of the conclusions is that 'Combinations of virtual and physical laboratories offer advantages that neither one can fully achieve by itself'.

The analysis coincides in part with Ma and Nickerson's work, as in *Science* it was stressed that virtual laboratories may facilitate conceptual comprehension, as they spare the student the sometimes complicated task of instrument handling. In this case, the authors insist that correct guidance during the remote experiment will greatly improve the student's experience.

More specifically, Marques *et al.* (2013) and García-Zubía *et al.* (2016) have analysed the impact of the VISIR remote laboratory on student learning. In the second case, the authors analysed via pre-test and post-test scores over 100 students' knowledge of basic electronics on two different campuses with two different teachers and with different prior experience (see Fig. 4.2). In both cases, the learning was statistically relevant, and overall the study established that the *d* value of the Cohen's size effect was one, which is an appreciable impact and 'similar to that of a teacher' (García-Zubía *et al.*, 2016). In other words, using VISIR in the classroom with students of electronics and electricity is a correct teaching option.

Following a different route and with a much larger sample (over 1000 students), Marques *et al.* (2013) observed that the students who had used VISIR during the course learned more than those who had not (see Fig. 4.3). The groups analysed were very varied in terms of subjects,

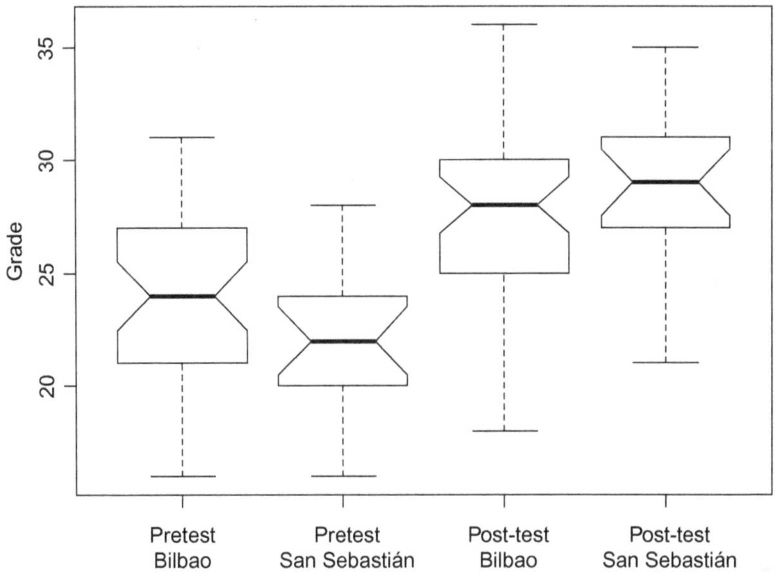

Figure 4.2. VISIR results in the learning process according to García-Zubía *et al.* (2016).

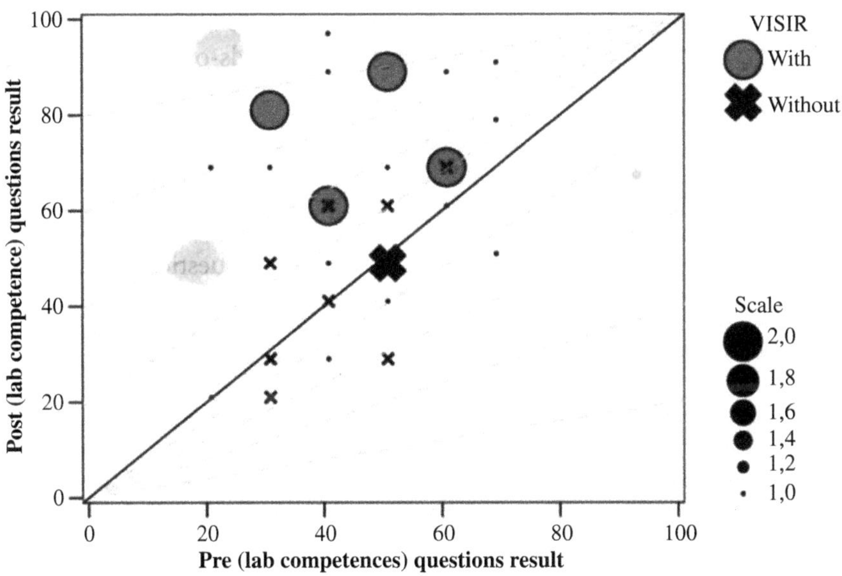

Figure 4.3. VISIR results in learning process according to Marques *et al.* (2013).

teachers, experience, etc., but nonetheless the use of VISIR had a positive impact upon the students.

Viegas *et al.* (2018) also analysed the use of VISIR by 417 students in Brazil within the framework of the VISIR+ project. The analysis reveals that 'teachers' involvement plus their ability to brief students on VISIR's usefulness have a significant influence not only on students' performance but also on their perception of learning and satisfaction with the tool'. This study also shows that the students recommend the modernisation of the VISIR's interface, but this interface is one of the most valued characteristics of VISIR from the teachers and researchers point of view.

One deduces from the previous paragraphs that remote experimentation does not offer worse learning results than classical experimentation, and that in some aspects it offers clear advantages. This conclusion is backed up, as far as possible, by specific data and studies, but why is remote experimentation positive in general? What elements assign this status? Where is the didactic value of a remote laboratory?

In the first three aforementioned studies, several of these elements stand out. First, evident in many cases is the fact that scientific instruments are already computerised and therefore when they are measured via the Internet the change is not so great. In other words, there is not so much difference between hands-on and remote experiments. Moreover, there is no need to dispense with either of the two, as both can be used. Even Ma and Nickerson (2006) recommend beginning with the hands-on and continuing with the remote experimentation: the former will provide security and the latter will allow for more agile work. de Jong, Linn and Zacharia (2013) stresses that what is most important is the guidance received by the student, in other words, the support we give to the investigation process performed by the student in the online laboratory.

For electronic circuits, Soysal (2000) asked this question: are remote laboratories recommended for assembling and measuring electronic circuits? His answer was no, and he reached it using the data given in Table 4.1, where a higher number of stars (*) denotes a higher value. The original table compared eight features, and in this case only five of them are reflected, the most relevant and illuminating. Beyond a detailed analysis, Soysal reveals that 'practices connected to a computer' is what he rates most highly as it combines the best of individual practice (total control over the experiment) with computerised instrumentation (quality and quantity of data collected). However, we see that his evaluation of 'WebLab' (remote laboratory) is very low and he only highlights the quality and quantity of

Table 4.1. Evaluation of different types of laboratory according to Soysal (2000).

	Individual projects	Guided practices in the laboratory	Computer simulations	Practices connected to a computer	Remote practices using WebLab	Remote practices using VISIR
Assembling the experiment	****	****	***	****	—	****
Instrument selection	****	**	—	**	—	***
Instrumentation	****	****	*	****	**	****
Real conditions of the experiment	****	****	—	****	—	****
Analysis of the results	**	**	****	****	****	****

data collected. What is more, 'computer simulation' is almost more highly rated than 'WebLab'. However, this table was created in 2000. At that time, Soysal's assessment was undoubtedly correct, but if VISIR is used as a remote laboratory for electronics, the evaluation should be indicated in the final column. VISIR allows the student to assemble and measure a circuit in a similar fashion to how it would be done in a hands-on laboratory, the difference being that instead of hands and eyes, they would use the interface and the mouse.

Table 4.1 illustrates how important it is for each instructor or educational centre to evaluate the suitability of each remote experiment, and not of all in general. Moreover, this evaluation may depend to a large degree upon the instructor and their way of teaching. What is important is that each instructor should perform this exercise of analysis, as they would for any other educational tool. For instance, Nedic *et al.* (2003) made a similar analysis but with other results. In their work, the main topic was Traditional and Alternative Methods of Performing Experimental Work, and the results were obtained for Real Lab, Virtual Lab and Remote Lab.

However, there is one question: is there a general and non-specific way of evaluating a specific remote laboratory or these laboratories as whole? ABET (Accreditation Board of Engineering and Technology, https://www.abet.org/) addresses in rigorous fashion the description of the competences particular to engineering (Feisel and Peterson, 2002) and (Feisel and Rosa, 2005), devoting a chapter to the specific competences of the laboratory. So, the question might be: does a remote laboratory fulfil the 13 learning objectives of ABET? (see Table 4.2).

The first nine learning objectives clearly depend on the features of the laboratory, remote in this case, with the other four excluded from this laboratory. And five of these nine stand out (OB1, OB5, OB6, OB8 and OB9) because they very much depend on the features of the remote laboratory. According to ABET, which is a standard recognised and used by different educational centres, a remote experiment or set thereof can be categorised according to the degree to which they help the student to fulfil these objectives or depending on the extent to which a remote laboratory hinders their fulfilment. It is also important to compare the remote laboratory with the corresponding hands-on version since not all hands-on laboratories are the same. Once again, this is a task that each instructor can perform in order to know the potential and limitations of their chosen laboratory.

For example, Psychomotor, OB8, is a learning objective not fulfilled by a remote laboratory by its very nature. Although in this case it should be

Table 4.2. ABET learning objectives by Feisel & Rosa (2005).

Objectives	Description
OB1: Instrumentation	Apply appropriate sensors, instrumentation, and/or software tools to make measurements of physical quantities.
OB2: Models	Identify the strengths and limitations of theoretical models as predictors of real world behaviours. This may include evaluating whether a theory adequately describes a physical event and establishing or validating a relationship between measured data and underlying physical principles.
OB3: Experiment	Devise an experimental approach, specify appropriate equipment and procedures, implement these procedures, and interpret the resulting data to characterize an engineering material, component, or system.
OB4: Data Analysis	Demonstrate the ability to collect, analyze, and interpret data, and to form and support conclusions. Make order of magnitude judgements, and know measurement unit systems and conversions.
OB5: Design	Design, build, or assemble a part, product, or system, including using specific methodologies, equipment, or materials; meeting client requirements; developing system specifications from requirements; and testing and debugging a prototype, system, or process using appropriate tools to satisfy requirements.
OB6: Learn from Failure	Recognize unsuccessful outcomes due to faulty equipment, parts, code, construction, process, or design, and then re-engineer effective solutions.
OB7: Creativity	Demonstrate appropriate levels of independent thought, creativity, and capability in real world.
OB8: Psychomotor	Demonstrate competence in selection, modification, and operation of appropriate engineering tools and resources.
OB9: Safety	Recognize health, safety, and environmental issues related to technological processes and activities, and deal with them responsibly.
OB10: Communication	Communicate effectively about laboratory work with a specific audience, both orally and in writing, at levels ranging from executive summaries to comprehensive technical reports.
OB11: Teamwork	Work effectively in teams, including structure individual and joint accountability; assign roles, responsibilities, and tasks; monitor progress; meet deadlines; and integrate individual contributions into a final deliverable.
OB12: Ethics in the Lab	Behave with highest ethical standards, including reporting information objectively and interacting with integrity.
OB13: Sensory Awareness	Use the human senses to gather information and to make sound engineering judgements in formulating conclusions about real-world problems.

said that the remote laboratory develops a similar objective but of a digital nature, it is a digital skill. The same thing happens with OB9, Safety. Can safety standards be integrated within the remote laboratory? With what degree of control by the student? The answer is no, although this question is also pertinent to a lesser extent in the classical laboratory.

For instance, for OB8 and OB9, the VISIR allows students to make mistakes, does not autocorrect them and allows them to create any circuit using a finite set of devices and instruments. This limitation may appear to be an impediment to the development of student creativity, but can they do this in a hands-on laboratory? Can a student select any resistor and assemble any circuit? In this, case the instructor usually chooses part of the material and the instruments, just as in the remote laboratory. Or if we imagine a programmable robot, in principle in a remote laboratory, students can perform the same practices, but cannot assemble the robot with their own hands. But do they usually do this in a normal laboratory or is it already assembled?

To summarise, any analysis of a remote laboratory can be undertaken with an educational approach and using statistical tools, or qualitatively, using ABET criteria or the criteria that seem relevant to the instructor, always taking into account that the remote laboratory can be analysed by itself, or compared with the classical laboratory.

The Table 4.3 relates some remote experiments to the first nine ABET criteria. An H (high) means that the remote experiment clearly achieves this criterion, M means medium and L means low. Once again, it is vital for each instructor to perform the corresponding analysis for each experiment and its implementation, whether it be remote or hands-on. A good exercise is for the reader to find and classify new remote experiments, even using different criteria.

A similar analysis is performed in (Zvacek and Restivo, 2018). In this case, the authors link an OE@FEUP remote experiment with a set of educational criteria such as Student Attention, Objectives, Recall of Prior Learning, etc. to analyse its strengths and weaknesses and recommend a series of guidelines in order to improve the educational performance of the remote experiment.

4.2 Use Cases of Remote Experiments in the Classroom

As has already been noted, remote experimentation enables students to participate in scientific and technological experiences without having to go to a real laboratory and without having to use simulators, occupying a place between both extremes. But how are remote laboratories used in the classroom?

Table 4.3. Analysis of different remote experiments according to ABET objectives.

	OB1	OB2	OB3	OB4	OB5	OB6	OB7	OB8	OB9	TOTAL
Archimedes LabsLand	L	H	M	M	M	L	L	L	L	L–M
Pendulum LabsLand	M	H	L	H	L	L	L	L	L	L–M
Radioactivity LabsLand	L	H	M	H	M	L	L	L	L	M
Planarias LabsLand	M	M	L	M	L	L	L	L	L	L
VISIR	H	H	H	H	H	H	M	M	H	H
Arduino robot	n.a.	H	M	H	M	H	H	M	H	M–H
Controller UNILabs	L	H	M	H	M	M	L	L	L	M
Controller OE@FEUP	L	M	L	H	L	L	L	L	L	L
Pendulum e-lab	L	H	L	H	L	L	L	L	L	L
Circuits LaboRem	L	H	M	H	M	M	M	M	M	M

Notes: OB1: Understood as the student's capacity to sensorise an experiment (H) or to access direct information from an experiment (M). If the interface delivers the information, then it is L.
OB2: Understood as the student's capacity to compare the experiment with the underlying model or algorithm
OB3: Understood as the student's capacity to create on their own (H) or have access to a sufficient number of experiences (M).
OB4: Understood as the possibility the interface offers the student of interpreting data via graphs (H) or simple data (M).
OB5: Equivalent to OB3 in a remote experiment.
OB6: Understood as the possibility the experiment offers the student of making a mistake and learning from it.
OB7: Understood as the student's capacity to be creative with the remote experiment.
OB8: Understood as the remote experiment's capacity to develop the student's digital psychomotor skills.
OB9: Understood as the remote experiment's capacity to present the student with real security problems.

Not so long ago, classroom work was focused above all on teaching, teachers were at the centre of the process, and textbooks were their support and reference. Assessment was via exams. Now one speaks more of teaching-learning, where the student plays an important role, so it is not enough for teachers to teach: the student should learn actively, should form part of the subject. This change of paradigm, from teacher-centric to student-centric, is reinforced or clarified by concepts like investigation, PBL, flipped classroom, etc. This change has coincided in time with the Bologna Declaration, which, although initially only had a political impact, has also been used as a lever for a methodological and pedagogic change, with a move from subject content to students' competences. It does not make much sense to discuss previous strategies in this text but to see how remote laboratories can be used to address prior challenges.

Let us consider a basic electronics lesson in secondary school or buoyancy in primary school. The teacher normally has to choose between three options: (1) first explain the model or behaviour of the system and then check this in the laboratory; (2) first go to the laboratory and there discover the model or behaviour of the system; or (3) simply dispense with the laboratory. The first is the classic approach, the second encourages students to investigate and the third omits the students' practical experience. During a course or subject, a teacher can mix the three approaches.

The above is expressed from the point of view of the teacher who is going to use the remote experiment in the classroom, but one also needs to bear in mind the designer, who is often a teacher too. When designing the remote experiment, the teacher/designer must opt for one of the following approaches:

- The experiment is bare, without teaching material, so any other teacher with knowledge can use it in the classroom with their students.
- The experiment has teaching material that may or may not be used by the teacher in the classroom.
- The experiment is included within teaching material so the two cannot be separated and the teacher has to use everything in the classroom.

In the last case, the teacher/designer has clearly been very generous, but this might not attract many teachers, because the material is too complex or too simple, or very extensive, or for any other reason. The balance between the experiment and its teaching material may have a crucial influence upon how much an experiment is used, irrespective of its technical complexity, etc.

We will now describe various cases of use of remote experiments in classrooms at different levels and always from an investigative perspective. Most of the laboratories have already been described, and all are offered by Labsland. Remote laboratories from other providers are not described as there is no guarantee that they can be accessed in the long term.

4.2.1 *Buoyancy in Archimedes' Laboratory*

Primary objective:
Discover why some objects float and others do not.
Understand the concept of density.

Secondary objectives:
Scientific method, data collection and graphic representation of data.

Context or introduction:
Some objects float — a football or a steel ship — while others do not — a marble or a huge stone — why?

Investigation:
The student launches objects and observes whether or not they float to discover why they float or under which circumstances. Thus, before launching a new object, they can guess whether or not it will float.

Procedure:

1. Note down an object's volume and mass.
2. Launch the object and see whether or not it floats.
3. Graph the results.
4. Explain or model the buoyancy of an object. Concept of density.
5. Evaluation.

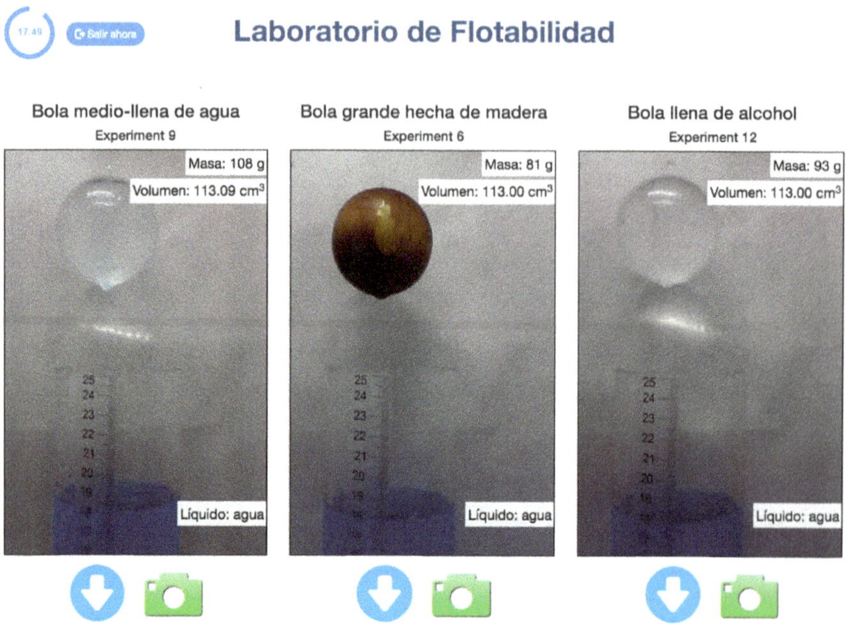

Figure 4.4. Example of use of LabsLand's buoyancy remote experiment.
Source: www.labsland.com.

Material available:
13 objects of different materials (glass, cork, wood, etc.) and densities that can be launched or contained in 13 test tubes with blue-coloured water (see Fig. 4.4).

Procedure walk through:

1. **Note down the volume and mass of a number of objects (not all of them, 8 for example).**

Object	Mass	Volume	Sinks	Floats
Experiment 1: Ball almost full of water	108.9 g	113.09 cm^3	—	—
Experiment 2: Large wooden ball	81.4 g	113.09 cm^3	—	—
Experiment 3: Ball filled with alcohol	93.2 g	113.09 cm^3	—	—
Experiment 4: Ball filled with marbles	135 g	113.09 cm^3	—	—
Experiment 5: Ball half-filled with marbles	—	—	—	—
Experiment 6: Ball with quite a few marbles	—	—	—	—

(*Continued*)

Object	Mass	Volume	Sinks	Floats
Experiment 7: Ball filled with glycerine	—	—	—	—
Experiment 8: Ball filled with water	—	—	—	—

2. Launch the objects and see whether or not they float.

Complete the table above with the results of the experiments

Object	Mass	Volume	Sinks	Floats
Experiment 1: Ball almost filled with water	108.9 g	113.09 cm^3	—	X
Experiment 2: Large wooden ball	81.4 g	113.09 cm^3	—	X
Experiment 3: Ball filled with alcohol	93.2 g	113.09 cm^3	—	X
Experiment 4: Ball filled with marbles	135 g	113.09 cm^3	X	—
Experiment 5: Ball half-filled with marbles	67.4 g	113.09 cm^3	—	X
Experiment 6: Ball with quite a few marbles	97.1 g	113.09 cm^3	—	X
Experiment 7: Ball filled with glycerine	140.2 g	113.09 cm^3	X	—
Experiment 8: Ball filled with water	113.9 g	113.09 cm^3	—	X

3. Graph the results.

Locate each object in the diagram, write its number inside and paint it green if it floats and red if it sinks (see Fig. 4.5).

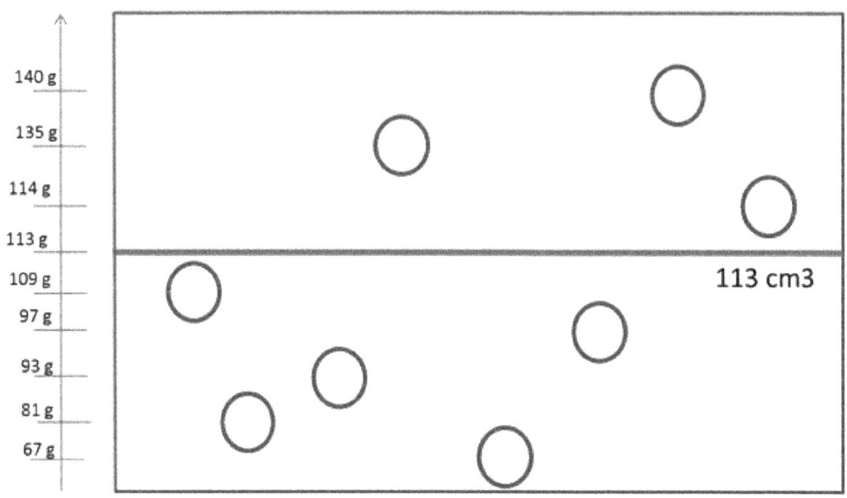

Figure 4.5. Graphic reflection of data in the buoyancy remote experiment.

4. **Explain or model the buoyancy of an object. Concept of density.**

   According to the previous graph, where are the objects that sink? Above or below the line? And the ones that float?

   Can you express what you see in words? Via an arithmetic operation?

   Here, each teacher will decide how to introduce the concept of density. It should not be forgotten that the remote experiment does not replace teachers, but simply accompanies their way of teaching.

5. **Evaluation.**

   In order to evaluate, the teacher has various options: exam questions, test, etc. But one way within the remote laboratory itself is by telling students to note down the volume and mass of the five remaining objects and predict whether they think they will float or not (hypothesis) before launching the five objects and seeing if the reasoning is correct or not.

Object	Mass	Volume	Sinks	Floats
Experiment 9: Large ball with marbles	135 g	6 cm^3	—	—
Experiment 10: Large wooden ball	—	—	—	—
Experiment 11: Ball filled with alcohol	—	—	—	—
Experiment 12: Ball filled with marbles	—	—	—	—
Experiment 13: Ball half-filled with marbles	—	—	—	—

Comments:
This remote experiment is pre-recorded and enables various students to connect at the same time, with the possibility of sharing results or collaborating actively, both inside and outside the class. Teacher and students can also work at the same time, making the guided investigation real.

In summary, this laboratory in no way conditions the use that can be made of it, but simply offers the possibility of launching objects into water to see whether or not they float. This kind of 'bare' experiment compares with others that not only offer the experiment but where the latter is integrated within a lesson from which it cannot be separated. Obviously, one kind is not better than the other. The former gives teachers more freedom and demands of them more work to situate the experiment in their classrooms, while the latter does not give teachers freedom, but everything is ready to be used without their intervention.

Additional experiments:
Apart from buoyancy, with this remote experiment it is possible to study relative density and pushing force.

In the first case, the experiment offers more than 13 objects. The new ones are launched into two different liquids apart from water: oil and sugared water. Thus, the student will see which objects that previously sank, now float, and vice versa.

In the second case, the student can see that the weight of an object changes in water and that the water level changes too. Via careful observation of this experiment, it is possible to measure the pushing force.

Furthermore, and provided the experiment is 'bare', the teacher can invent new practices with it.

4.2.2 *Calculation, assembly and measurement of circuits with resistors with VISIR*

Primary objective:
Discover the mathematical model of connection between resistors via experimentation.

Secondary experimentation:
Scientific method, measurement error and mathematical model.

Context or introduction:
A resistor is part of basic electronic circuits and has a value, which may range from just a few ohms (Ω) to millions of ohms (MΩ). Resistors can be connected in series or in parallel, creating a circuit with a corresponding total value in ohms. In Fig. 4.6, the connection R1, R2, … , Rn ultimately behaves like one single and total resistor between A and B.

Investigation:
Whenever students can assemble and measure connections between resistors, can they develop or estimate the mathematical model of this

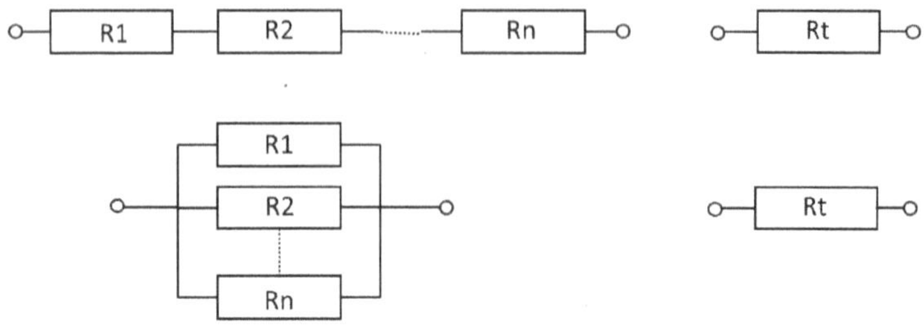

Figure 4.6. Resistors in series and in parallel.

connection? Can they calculate *a priori* the value of the total resistor of an assembly without assembling or measuring it?

Procedure:

1. Measure a single resistor and see that the value is not correct; that there is an error.
2. Assemble and measure connections with resistors in parallel and in series.
3. Make a note of the values and share them with colleagues.
4. Obtain the model of the connection in series.
5. Obtain the model of the connection in parallel.
6. Estimate or calculate the value of the total resistor of a series-parallel connection.
7. Evaluation.

Material available:
Four resistors, two of 1 kΩ and two of 10 kΩ, combinable in any way.
Protoboard to assemble the circuits.
Multimeter or tester to measure.

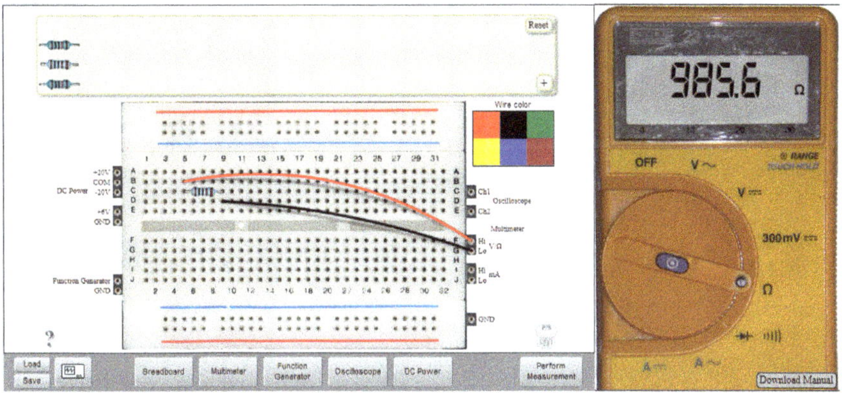

Figure 4.7. Resistance measurement in VISIR.

Source: www.labsland.com.

Procedure walk through:

1. **Measure a resistor.**

   Figure 4.7 shows the measurement of a resistor of 1 kΩ. We see that the value of the tester is not 1000 Ω, but 985.6 Ω. Therefore, there is an error of 14.4 Ω, or 1.44%. The error should be understood as a deviation between what is expected and what is obtained: the resistance is not 1000 Ω, the tester does not function correctly, etc. Is there exactly 1000 cm^3 in the milk carton? What does 'exactly' mean?

2. **Assemble and measure connections with resistors in parallel and in series.**

   Figure 4.8 shows the result of connecting in series and measuring two resistors of 1 kΩ and three resistors: two of 1 kΩ and one of 10 kΩ.

3. **Make a note of the values and share them with classmates.**

   It is a good idea to create a document on Google Sheets where students keep a record of the results obtained. Thus, each student

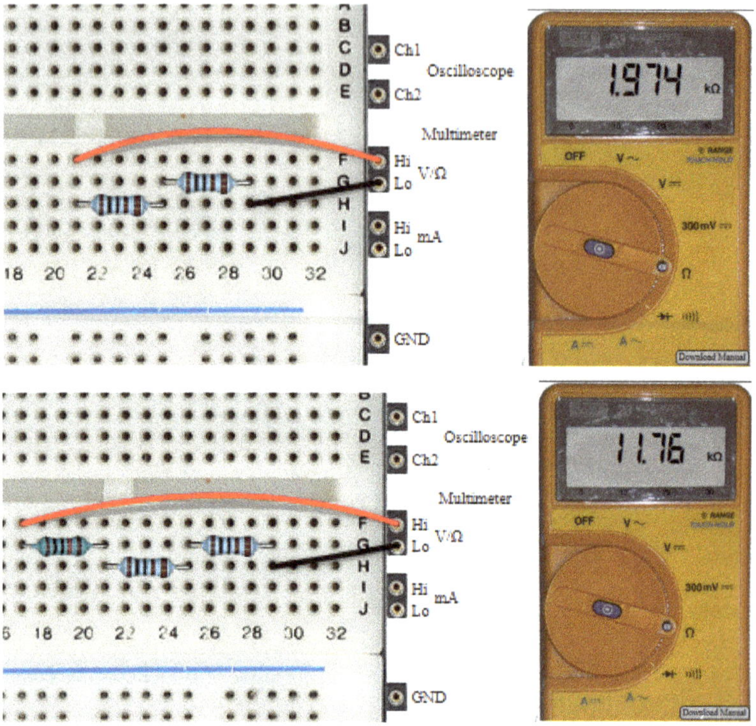

Figure 4.8. Measurement of resistors in series in VISIR.

Source: www.labsland.com.

sees if their results are the same as those of the rest, and the teacher can speak of mean value, etc.

4. **Obtain the model of the connection in series.**

   In view of the numbers, what is the mathematical model that predicts the value of the connection in series of four resistors: two of 1 kΩ and two of 10 kΩ?

   Solution: Mathematical model in series, Rtotal = 1+1+10+10 = 22 kΩ.

   Can you check if the model is correct?

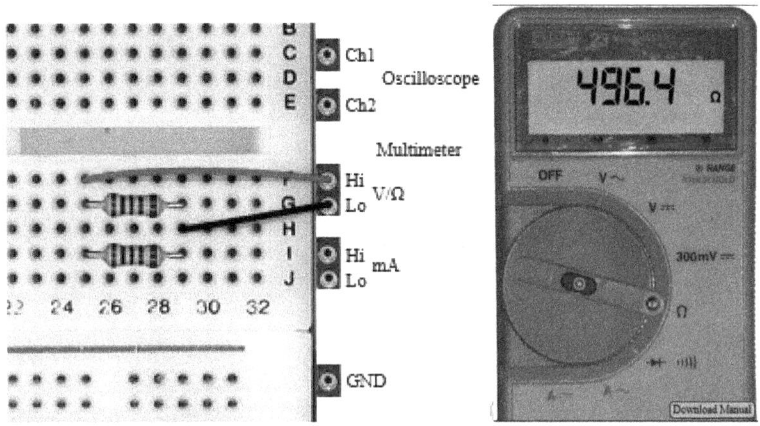

Figure 4.9. Measurement of resistors in parallel in VISIR.

Source: www.labsland.com.

5. **Obtain the model of the connection in parallel.**

   If the assembly is of two resistors of 1 kΩ in parallel, what is the value of the total resistor? The best thing in this case is to assemble and measure first, and then think as is shown in Fig. 4.9.

   The process of assembling and measuring resistors in parallel is quite simple, although its interpretation is more difficult. What mathematical model calculates the connection of various resistors in parallel? Perhaps in this case the teacher can give them the model, see Fig. 4.10, and the students check whether it is correct, or look for it online.

6. **Estimate or calculate the value of the total resistor of a series-parallel connection.**

   What is the approximate value of the connection in Fig. 4.11? Can you estimate the value of the total resistor?

   The student can estimate a value and then assemble and measure to see whether or not the estimation was correct.

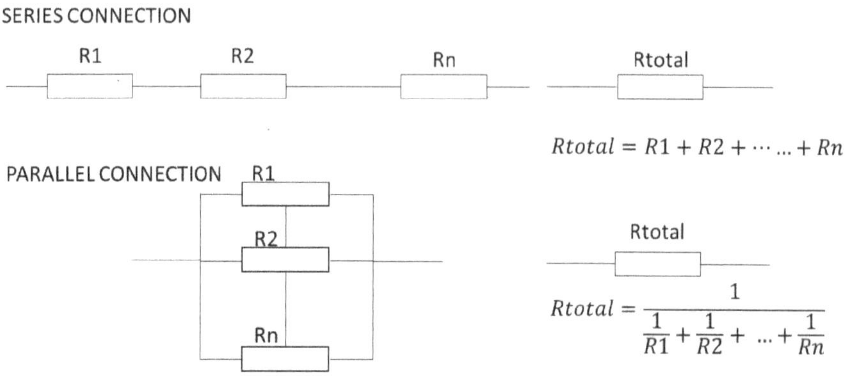

Figure 4.10. Associations of resistors and mathematical models.

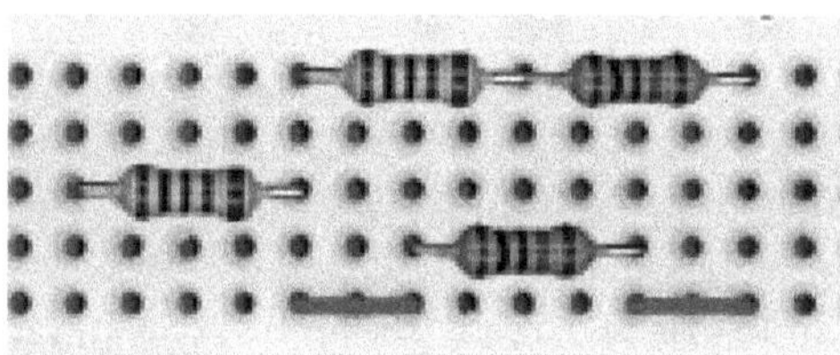

Figure 4.11. Association of series-parallel resistors.

Source: www.labsland.com.

7. **Evaluation.**

   The teacher can evaluate knowledge in various ways: via new connections, via a test, via a task, etc.

   Or pose the challenge of Fig. 4.12: Can you design a connection of resistors with an approximate value of 1.95 kΩ using only two resistors of 1 kΩ and two of 10 kΩ?

Comments:
The model proposed an investigative approach, but clearly the teacher can opt to use VISIR to test the model: first calculate and then assemble and measure to thus test the goodness of the model. VISIR is agnostic *vis-à-vis* the teaching approach, it simply makes it possible to assemble and measure. The rest is the responsibility of the teacher.

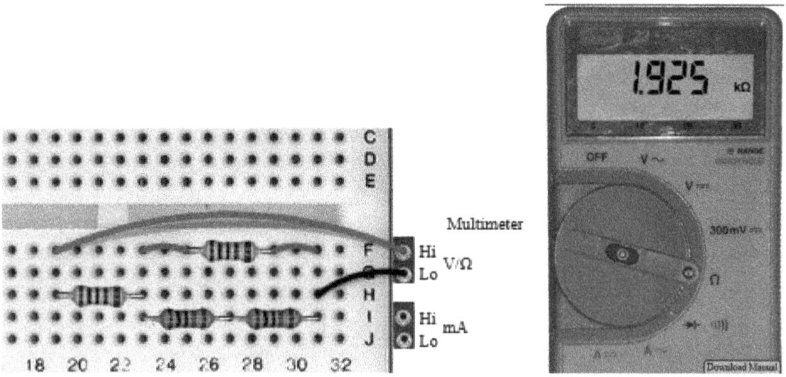

Figure 4.12. Assembly and measurement of association of resistors in VISIR.
Source: www.labsland.com.

Any teacher in this field knows that teaching how to assemble and measure resistors in a hands-on laboratory has its complications. VISIR makes it possible in a very short time for the student to be able to assemble and measure, as they learn directly from the teacher's example on the board, all the more so if the classroom has a digital board.

In the VISIR, but not in other remote experiments, various users can work at the same time, each on their own circuit. In other words, the teacher can show the students how to work, and they can do so at the same time as the teacher. In this way, the experience is more interactive, breaking the pattern of the teacher does and the student sees. This also breaks the pattern of teacher explains today in class and students go to the laboratory in a few days' time.

Moreover, the student can do the same work from home as in the classroom, or can share the experience with their family. Remote experimentation is an innovative tool in the teacher's hands.

VISIR also has a dashboard that offers the teacher many possibilities, as is explained in Section 7.1.

Additional experiments:
In the next level, the circuits will be powered with direct current and therefore it will be possible to measure the current and the voltage in each of the resistors. Once students master these measurements, they will be able to test Ohm's and Kirchhoff's Laws. The teacher will then be able to set the students design exercises. When designing, the student should ensure that

the circuit in question fulfils certain voltage and/or current requirements. The VISIR will suffice to analyse and design.

4.2.3 *Pendulum and pendular movement*

Primary objective:
Understand the movement of a pendulum and the variables that affect it.

Secondary objectives:
Scientific method, signal measurement and misconceptions.

Context or introduction:
A pendulum has two parts: a rod and a weight at one end of the rod, with the other end attached to an axis. If we raise the pendulum and release it, it starts to move to the right and to the left, repeating the movement. What does the speed of the movement depend upon?

Investigation:
The pendulum clock ticks at the speed marked by the pendulum; if this is not correct, then the clock will be fast or slow. How can this be corrected?

Procedure:

1. Raise the pendulum at a specific angle.
2. Launch and measure the movement of the pendulum.
3. Order the results for all the angles.
4. Evaluation.

Material available:
A pendulum and an interface to control the angle of release and measure the movement.

Procedure walk through:

1. **Raise the pendulum at a specific angle.**

   Decide what angles are going to be measured and start with the first, 10° (see Fig. 4.13).

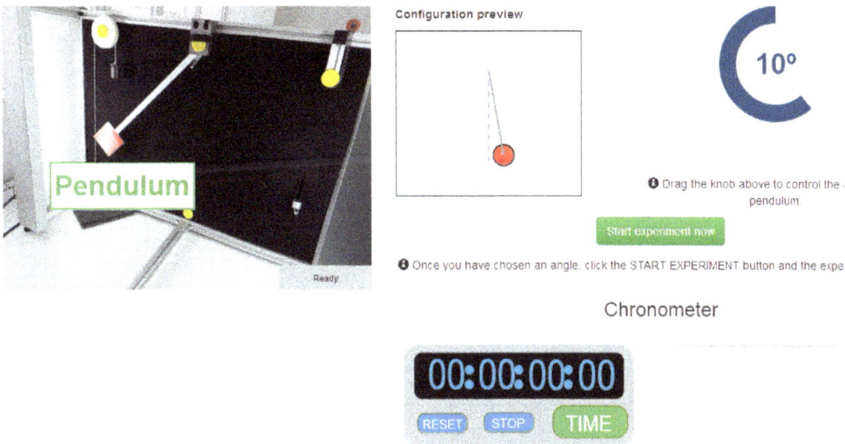

Figure 4.13. Interface of LabsLand pendulum remote experiment.
Source: www.labsland.com.

2. **Launch and measure the movement of the pendulum.**

When the pendulum is released, it begins to oscillate from right to left, and back again. To measure the movement a period of time is used: how many oscillations per second, or how long an oscillation lasts.

The interface offers at least four ways of measuring the period:

- Start the chronometer and count the oscillations produced in 10 seconds, for example. Period = no. oscillations/10.
- Start the chronometer and begin to count oscillations up to five, for instance, and stop the clock. Period = 5/time in seconds.
- Measure on the graph the time lapse between five complete oscillations, for instance. Period = 5/time in seconds.
- Download the Excel file and measure the period reading the detailed data.

The first two look easy, but ultimately they require concentration. The last two are more accurate but less fun. In order to achieve a certain quality of measurement, it will be necessary to repeat several times.

3. Order the results for all the angles.

One simply needs to note down the period calculated for each angle. Once again, this is a task that can be shared among other students via Google Sheets.

	10°	20°	25°	30°	35°	40°	45°
Method 1	—	—	—	—	—	—	—
Method 2	—	—	—	—	—	—	—
Method 3	—	—	—	—	—	—	—
Method 4	—	—	—	—	—	—	—

4. Evaluation.

The first question to be answered is: does the pendulum period depend on the angle of release? Why?

According to the theory, the period of a pendulum is independent of the angle and only depends (roughly speaking) on length according to the following formula

$$T = 2 \cdot \pi \sqrt{\frac{l}{g}}$$

Knowing that the length is 45 cm, and using the formula, can you calculate the value of *g*, the known value of which is 9.8 m/s^2? What value for g do you obtain? With which method of measurement of the period do you obtain a more accurate value for *g*?

The value of the period of this pendulum is constant, but not completely for every angle. What angle provides the most accurate value for *g*? For which angle is the aforementioned mathematical model most accurate? For high or low angles?

Comments:
The pendulum remote experiment is pre-recorded, so all the students in the class can simultaneously access the experiment along with the teacher. Consequently, the development of the experiment can be shared and in real time; in other words, sharing results and measurements as they are obtained.

4.2.4 *Radioactivity and effects on health*

Primary objective:
Identify different types of particles, their radioactivity and their effect on health.

Secondary objectives:
Scientific method, measurement of variable signals and graphic representation of signals.

Context or introduction:
Materials emit radioactivity (more or less) spontaneously, and in doing so emit alpha and beta particles and gamma and *X* radioactivity.

Figure 4.14 shows how the different particles and the radioactivity behave if they have to pass through different materials.

The alpha particle is large (an atomic nucleus), very hazardous and easily stopped.

The beta particle is smaller (an electron), less hazardous and not easily stopped.

Gamma rays are very hazardous and pass through all kinds of materials.

Investigation:
In 2006, Litvinenko was assassinated by the use of polonium diluted in tea. What kind of deadly radioactive particles are emitted by polonium? Alpha, beta or gamma?

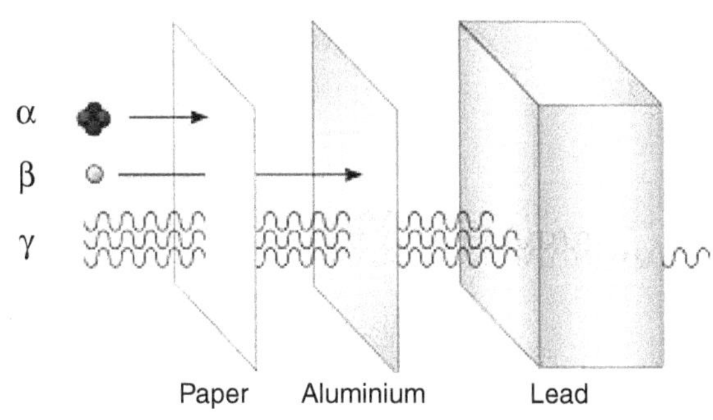

Figure 4.14. Radioactivity, particles and absorbers.

Source: https://en.wikipedia.org/wiki/Radiation.

Procedure:

1. Measure the radioactivity of different materials via the number of impacts of their particles and radioactivity.

2. Repeat the previous step after fitting different absorbent materials.

3. Match each type of radioactivity with a different material.

4. Evaluation.

Material available:
Three types of materials — americium, cobalt and strontium — and different kinds of absorbers: paper, aluminium and lead. In addition, the bombardment can be measured from different distances and over different lengths of time.

Procedure walk through:

1. **Measure the radioactivity of different materials through the number of impacts of their particles and radioactivity.**

   We measure the number of impacts for each material without the absorber for five seconds at a distance of 15 cm. These last two parameters remain unchanged in all the experiments (scientific method).

   For example, Fig. 4.15 shows us the experiment for strontium.

   Observing the results in Fig. 4.16, we can see that Radioactivity-1 has recorded around 225 impacts (a maximum of 250, a minimum of 200, so around 225 impacts). Students can share their results on Google Sheets in order to have a greater wealth of information.

2. **Repeat the above after fitting different absorbent materials.**

   In total, there should be 12 measurements, and each lasts several minutes. Once again, it may be a good idea for students to complete the data in collaboration.

Figure 4.15. LabsLand radioactivity remote experiment located at the U. of Queensland. *Source*: www.labsland.com.

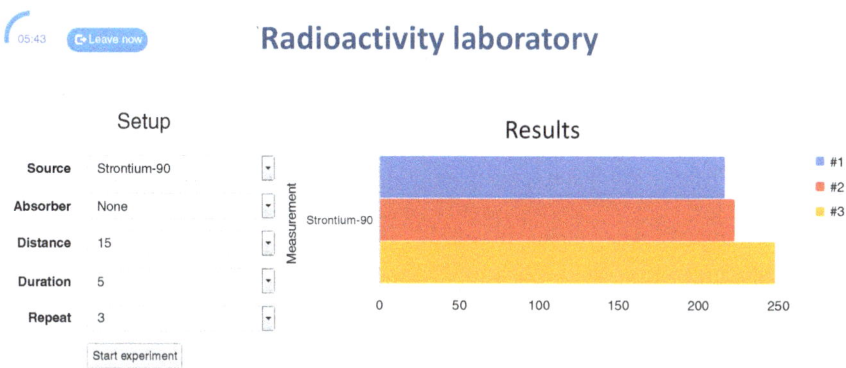

Figure 4.16. Interface of results of LabsLand radioactivity remote experiment. *Source*: www.labsland.com.

	With nothing	With paper	With aluminium disc	With lead disc
Strontium	250	225	—	—
Americium	—	—	—	—
Cobalt	—	—	—	—

Each student should write inside each circle the number of impacts (see Fig. 4.17).

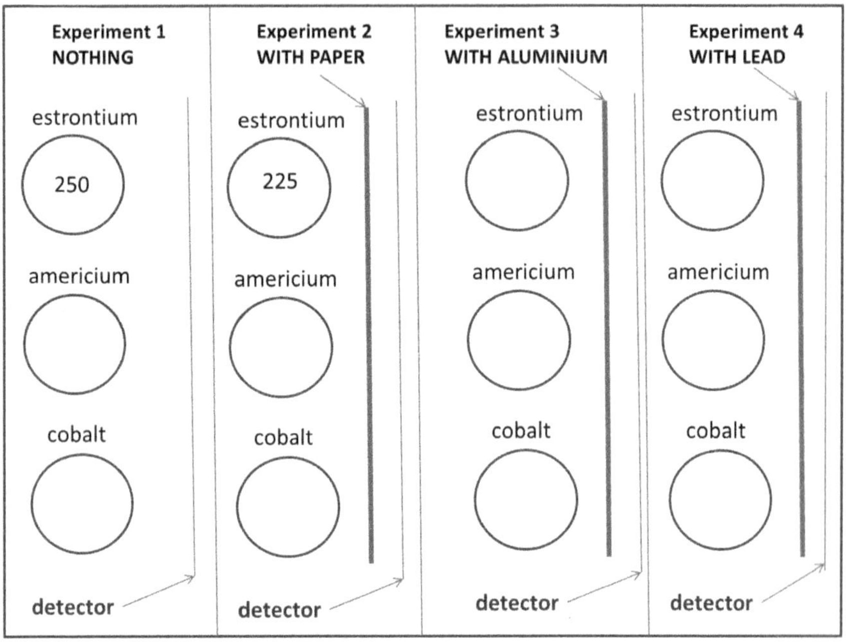

Figure 4.17. Graph to show results of radioactivity experiment.

3. Match each type of material with a type of radioactivity.

The number of impacts of alpha particles should fall sharply with just a sheet of paper, and be reduced to zero or thereabouts with the lead disc.

Therefore, and for example, strontium does not emit alpha particles, as the number of impacts has barely fallen with paper. In other words, the paper has stopped hardly any particles.

4. Evaluation.

Evaluation simply involves each student saying what kind of particles are emitted by each material.

They should then think about what kind of particles were emitted by the polonium to kill Litvinenko.

Another two questions are: why did the assassin not die, and why did they dilute the polonium in tea? Was it not sufficient simply to place it in front of Litvinenko in order to kill him?

Comments:
The radioactivity remote experiment is pre-recorded, so all the students in the class can simultaneously access the experiment along with the teacher. Therefore, the development of the experiment can be shared and in real time, in other words, testing the hypotheses as they arise.

Additional experiments:
In addition to the previous activity, it is interesting to see how the variables of distance and time affect the number of impacts. If distance is halved, do the number of impacts double? Is it a linear relationship?

This is a typical example of scientific experiment. There are four variables: material, absorber, distance and time. Three are fixed and one varies. Only one variable should ever vary at a time, otherwise we would not know to which of them corresponds the anticipated effect.

Consequently, we decide always to use strontium without absorber. In the first case, we set a distance of 15 cm and we assign various values to the time. Then we repeat this, but setting the time at five seconds and varying the distance.

4.2.5 *Planarians and stimulant beverages*

Primary objective:
Identify the effects upon the nervous system of the consumption of stimulant beverages.

Secondary objectives:
Biological experiment, scientific experiment, signal measurement and graphic representation of data.

Context or introduction:
The planarian is a tiny living aquatic organism that has a 'similar' nervous system to humans. So if we place a planarian in different liquids, we will see how these affect its behaviour.

Investigation:
Is it healthy to drink certain beverages without any type of control? Can they affect our behaviour?

Procedure:

1. Choose a beverage.
2. Measure the planarian's activity.
3. Order the results for all the beverages.
4. Evaluation.

Material available:
Samples that contain a planarian in five liquids: water, sugared water, ginseng, caffeine and taurine.
A chronometer and activity counter.

Procedure walk through:

1. **Choose a beverage.**

   The five beverages have to be chosen one by one (see Fig. 4.18).

2. **Measure the planarian's activity in the chosen liquid.**

   Having chosen water, we access the experiment interface, where we can see the sample, a chronometer (below the sample) and a counter, as shown in Fig. 4.19.

   The user measures activity during a specific period of time, for example, 30 seconds, clicking on count up every time the planarian moves. However, how do we measure activity? When the planarian crosses the line or when the whole planarian is in a new square or when ...?

   This is an interesting question in the classroom, as we need to find an indicator that is valid and easily obtainable. For instance, add one every time the planarian crosses a line, or each time the planarian moves to another square, etc. But when does it change squares? When the planarian's whole body does so or just the front part?

   Once again, sharing the experiment via Google Sheets is a good idea as this provides students with more data.

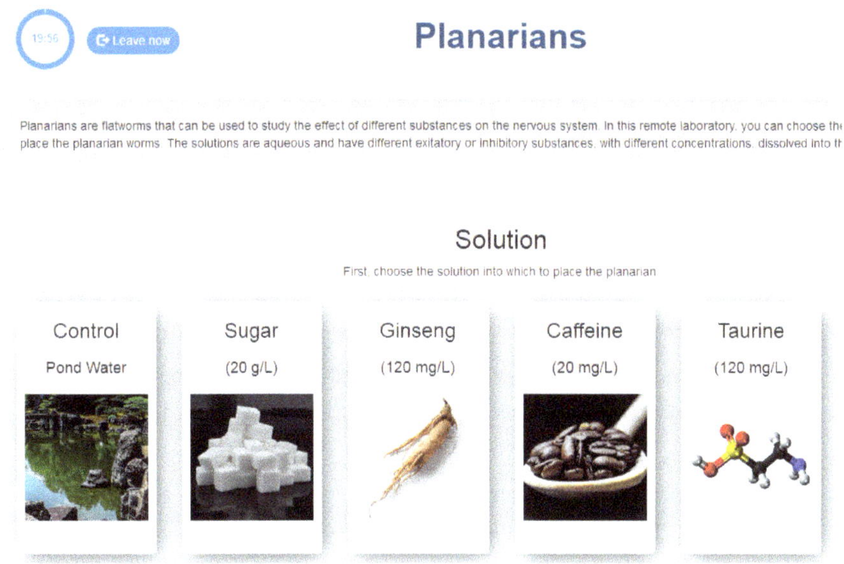

Figure 4.18. Interface of configuration of LabsLand planarians remote experiment.
Source: www.labsland.com.

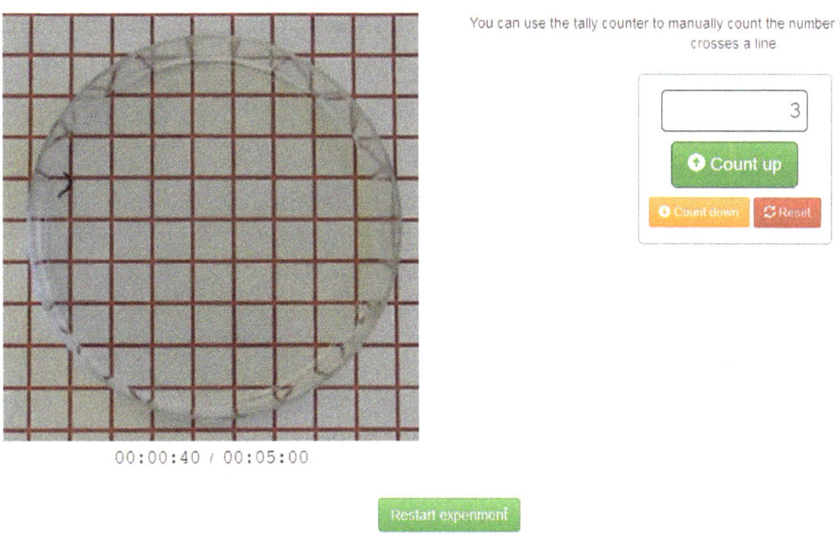

Figure 4.19. Interface of LabsLand planarians remote experiment.
Source: www.labsland.com.

3. **Order the results for all the beverages.**

   One simply needs to complete a table or find the graphic representation that facilitates analysis.

	Water	Sugared water	Water with ginseng	Water with caffeine	Water with taurine
Activity	—	—	—	—	—

4. **Evaluation.**

   This is as simple as ordering which products have the greatest effect and then checking which beverages contain these products, observing the proportions.

   Students should understand the lesson and apply it to their own lifestyles.

Comments:
The planarians remote experiment is pre-recorded, so all the students in the class can simultaneously access the experiment along with the teacher. Therefore, the development of the experiment cans be shared and in real time, in other words, analysis and conclusions can be undertaken together.

4.2.6 *Remote educational robotics*

Robotics is regarded as one of the most powerful tools to develop computational thinking, which in turn is one of the competencies required by all students in order to meet the demands of 21st century society.

LabsLand's ArduBlock remote experiment offers the student and the teacher the possibility of designing an algorithm using the Blockly graphic tool, subsequently to be tested with a real robot whose behaviour is observed via webcam. For example, on left of Fig. 4.20 we can see the algorithm code that moves the robot, and on the right the robot in motion.

In this case, neither teacher nor student are responsible for the assembly or maintenance of the robot, as the latter is located elsewhere, and the experience focuses on programming the robot. This does not mean that students cannot assemble their own robot in the classroom (the ArduBlock robot is commercial), but that they have access to a robot

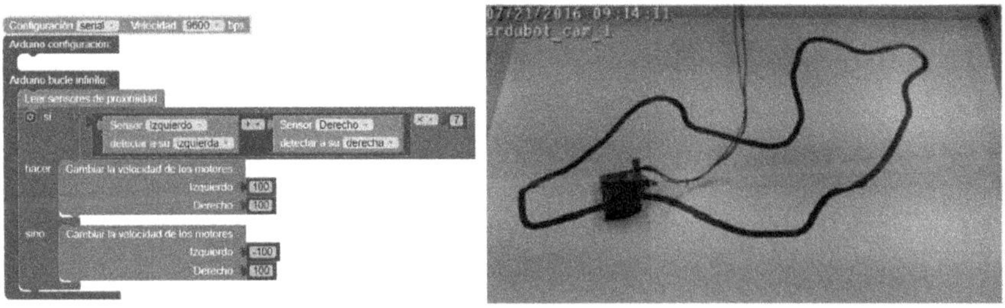

Figure 4.20. Program and remote experiment with LabsLand robot, Ardubot.
Source: www.labsland.com.

```
void calibrateSensors()
{
  // Wait 1 second and then begin automatic sensor calibration
  // by rotating in place to sweep the sensors
  delay(1000);
  for(uint16_t i = 0; i < 120; i++)
  {
    if (i > 30 && i <= 90)
    {
      motors.setSpeeds(-200, 200);
    }
    else
    {
      motors.setSpeeds(200, -200);
    }

    lineSensors.calibrate();
  }
  motors.setSpeeds(0, 0);
}
```

Figure 4.21. Ardubot program code.
Source: www.labsland.com.

that always functions correctly so as to test their algorithms. Maintenance of robots can be an absolute nightmare for students, and for the teacher in particular. In this case, the teacher concentrates on programming the robot and can choose between Blockly and C language (or Python, etc.) as is shown by Fig. 4.21.

Clearly, it is not possible simultaneously to connect various students to the robot, as the latter can only obey one algorithm at a time. Therefore, the laboratory needs to contain various identical copies of the same experiment in order to provide smooth service to the entire class. In this case, ArduBlock has four identical programmable robots, as well as a queue management service. Experiments completed by LabsLand show that with these four copies of the robot it is possible to accommodate a class of 25 students without excessive delays. Bear in mind that students do not all need to access at the same time, and that they rarely require more than two or three minutes to check that the algorithm functions. Once again, the fact that it is remote allows students to demonstrate their progress in robotics at home. Or in a similar fashion, it permits students at any centre to increase their knowledge of robotics by accessing ArduBlock.

Additional experiments:
What experiments can be performed in class? Apart from learning to think and programme an algorithm, which is the basic activity in educational robotics, the teacher can give students a programme so they can say what it does. Or can give them a programme asking them to say why it does not work and how to fix it. Or can present them with a challenge like bursting an inflated balloon on the track.

The ArduBlock track contains various Formula one circuits drawn with lines and is capable of measuring how long the robot takes to complete a lap. This way, students can compete against one another, or with another educational centre. In this case, the robot's success will not depend on the hardware design, but exclusively on the algorithm designed, and therefore the competition can be much closer.

Once again, if the teacher does not like the robot or prefers to remotise the one they use in class, perhaps a Lego Mindstorm, they must then assume this task, with the possibility of sharing that result with other teachers. The design work for remote experiments can be seen in Chapters 5 and 6.

The following examples describe projects and initiatives focussed on remote and virtual laboratories but are not specific examples.

4.2.7 *GoLab ecosystem*

The GoLab ecosystem (https://www.golabz.eu/) is the result of two projects financed by the EU: GoLab (FP7) and NextLab (H2020) and coordinated by the University of Twente (Holland) with over 15 partners.

In the GoLab ecosystem, the online laboratories (virtual and remote) are part of the ILS (Inquiry Learning Spaces) where students learn about a STEM concept from an investigative perspective in five phases:

Contextualisation, Conceptualisation, Experimentation, Conclusion and Discussion. The online laboratory is perhaps the central part of the ILS, but only one part.

GoLab offers the reader the opportunity to see the remote laboratory not only as something useful per se but as part of a more elaborated pedagogical discourse. Moreover, GoLab is a repository of remote laboratories that permits the reader to study them.

4.2.8 *ELLI project*

ELLI (Excellent Teaching and Learning in Engineering Sciences) was a collaborative project with partners from RWTH Aachen University, Ruhr-University Bochum and TU Dortmund University. Funded within the "Teaching Quality Pact" by the German Federal Government and States, and ran in two phases (2011–2016 and 2016–2020), the project aimed at continuously improving the teaching quality and studying conditions of engineering programs in Germany (see: https://www.elli-online.net/en).

The ELLI Project has over 20 million euros of funding, which makes it a reference in its area. In ELLI, once again, remote experimentation is one more tool en route to the final objective: improve student learning. And this requires the collaboration of engineers, instructors and teacher training schools to adapt technology to teaching, and vice versa. The second part of the project, ELLI2, also merges the concepts of Industry 4.0 and artificial intelligence. The results of this project are being put into practice in engineering and teacher training colleges.

In our case, the most important aspect of ELLI is that it regards remote experimentation as a tool to change and reinforce learning in one of the most developed countries in the world.

4.2.9 *e-LIVES*

The e-LIVES project (https://e-lives.eu) should be seen as a continuation of EOLES (http://www.eoles.eu/). In both cases, the idea is for the use of remote laboratories not only to increase the experimental potential of the countries of North Africa but for this use to stimulate curricular reorganisation of university studies in the countries involved under the concept of e-Engineering.

The concept of e-Engineering results from the concatenation of two previous ideas: e-learning and remote laboratories. Responding to a necessity felt by Morocco, Algeria and Tunisia in North Africa and Jordan in the Middle East, the e-LIVES project, a three-year Erasmus+ project financed by the European Union, aimed to make e-Engineering courses mainstream, part of the general offer of engineering courses of Universities worldwide. With that goal in mind, a set of documents was produced, detailing how universities may move through the different e-Engineering course design and development stages (building of a curriculum, getting the national accreditation, training teachers, creating contents,...), and how they may develop and implement suitable remote laboratories.

The e-LIVES remote experiments are focused on electronics and consist of a group of circuits that can be modified via sets of relays so the user may enjoy rich and complex practical experiences. Technologically speaking, e-LIVES is based on the use of remotely configurable electronic equipment under LXI and the use of relays under the RLMS WebLab-Deusto to ensure the federation of the resulting experiments.

The most important aspect of e-LIVES is that it seeks not only to implement remote experiments that improve subjects of university electronics degrees but also to be a catalyst of structural changes in education and an enhancer of relations between countries.

4.3 Educational Value of Remote Experiments

It is worth noting once again that a good experiment is not only the product of advanced technology but must also establish its credibility by means of the interface, support material and involvement of the teacher in the classroom. Furthermore, the use of a remote experiment should produce a measurable learning result. As stated by the Emerging Technologies in Engineering Education report (Vries *et al.*, 2017): 'However, technological adaption does not necessarily mean better performance.... It is evident that the impact cannot be measured by the fact whether or not technology is being used, but by how well the technology is used to support teaching and learning'.

In the first place, one should enquire the students' perception after using a remote experiment. *Did you enjoy the experience? Would you like to use a remote experiment again? Would you recommend it to other students?* These questions are addressed in Section 7.1.

For instance, during the first years of de WebLab-Deusto, 2004–2009, hundreds of students accessed different remote experiments over 10,000 times. At the end of each year, every student completed a Likert survey

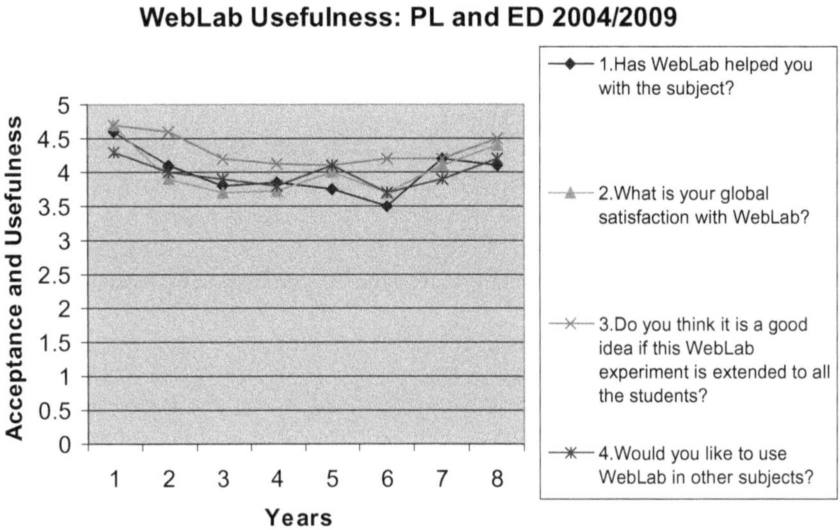

Figure 4.22. Results of the WebLab-Deusto usefulness survey.

on a scale of 1–5. Figure 4.22 shows the answers to some of the questions (Garcia-Zubía *et al.*, 2009b). It is interesting to note that throughout that period the students expressed a desire to use WebLab again, but were even keener for other students to do so.

However, it may be more interesting for teachers to ask themselves the following question: Can I perform a quantitative study of the effect of the use of a remote laboratory in my classroom? The most important piece of advice is that the reader should contact an expert in education, an educationalist or a psychologist, and also take into account from the beginning all the aspects related to ethics and data protection involved in research.

As an example, a research method is based on the use of pre-test and post-test. For instance, let us consider the pendulum experiment:

- The teacher prepares a test with a series of questions (pre-test) and gives it to the student to be completed before the class begins. The pre-test will probably not be very long (5–10 questions), and the student will have a lot of incorrect answers as they have not yet gone to class.

- The teacher organises the lesson in such a way that the remote experiment plays a fundamental role in the student's learning.
- The teacher prepares a post-test, which may be the same pre-test or another new one, and gives it to the student again after class.
- The results are analysed statistically and the appropriate conclusions are drawn.

Some explanatory comments with regard to the above process:

- The pre-test should not be long (5–10 questions is sufficient) and should not be complex, as it is not a case of finding out what the student knows in absolute terms, but how much has been learnt using the remote laboratory.
- The post-test can quite easily be the actual pre-test, perhaps reordering the questions.
- To reinforce the above process, one can employ a control group, in other words, other students will complete the same pre-test and post-test but without using the remote laboratory. Thus, one not only measures the effect of the remote laboratory but also compares it with other scenarios.
- The statistical process should not only contemplate mean and variance, but should follow a tried-and-tested method.
- The tests should be identified with students' names and surnames, which means following a protocol to guarantee data protection. This protocol depends on the investigative ethics committee of each centre or of a centre of reference. Students are not 'ours'.
- The question is can a teacher give an educational tool, like a remote laboratory, to some students but not to others?

The answer to this question and the protocol to be followed once again depend upon the investigative ethics committee.

There are many ways of evaluating the impact of a tool in the process and the learning results. All of these are documented and require the supervision of an expert in the subject and in education.

Part
3

Designers Point of View

This part describes the technical aspects of a remote laboratory from a designer's perspective, and in essence is extracted from the text prepared by Luis Rodríguez-Gil for the e-LIVES project financed by the EU (https://e-lives.eu/?page_id=10241#). Luis Rodríguez-Gil is co-founder and CTO (Chief Technology Officer) of LabsLand.

The objective is to guide the reader who wishes to design a remote experiment. What steps should they take? What technologies should they be familiar with? What support do they have? What can they rely on? If readers are defined simply as users of remote experimentation, they are likely to find this part complex and of little interest.

This part describes an architecture for educational remote laboratories. Although many potential architectures exist and many of them could lead to effective remote labs, a particular architecture is proposed here because it is well-proven and provides certain practical advantages for developing and maintaining the laboratory and guaranteeing its scalability and reliability. The selected architecture is WebLab-Deusto and is available at https://github.com/weblabdeusto. This open software is maintained by LabsLand (labsland.com).

Before addressing the design in detail, it makes sense to provide a conceptual description of the architecture of a remote laboratory. Remote laboratories have existed for over two decades (Nedic *et al.*, 2003) and (Gomes and Bosgoyan, 2009) they allow users to access equipment remotely through the Internet, and one of their main applications is education (Jong *et al.*, 2013). Remote laboratories comprise both hardware and software (Angulo *et al.*, 2018). Both parts are critical, and the importance of the software part of the system has been increasing (Nedic *et al.*, 2003) and (García-Zubía *et al.*, 2009b).

For a remote lab architecture to be able to lead to production-level labs effectively, it is of particular importance that it takes this fact into account. A common traditional issue with remote lab design and development is that engineers rarely have a strong background in both software and hardware development. This has often led to one of the two aspects being neglected in a remote lab (Angulo *et al.*, 2018). The effects of neglecting the hardware side are probably the most obvious, because the laboratory will simply not work, or the equipment will be unreliable, break often, and be hard to repair and maintain. The effects of neglecting the software side are sometimes less obvious but just as important, as this will often lead to poor reliability, poor maintainability, poor user experience, deployment issues, end-user bugs and security risks.

Therefore, the creation of a production-level remote laboratory will often require the collaboration of at least two different experts, one in hardware and one in software — unless the developer has high expertise in both fields. Additionally, for the remote laboratory to be actually useful, the collaboration of a pedagogical expert will also often be advisable, though that is beyond the scope of this Part 3.

The following figure shows the typical components of a web-based remote laboratory. The 'experiment hardware' is the hardware that is the target of the experimentation effort. For instance, in the case of an 'Arduino board' laboratory, this would be the actual Arduino board (and possibly sensors or inputs, if the laboratory includes them). The remotisation hardware is the additional components required to control the experiment hardware automatically (the motor in the pendulum). In the aforementioned case, for instance, that would be the hardware that programs the user-provided binary file (.bin file) into the Arduino board, and the hardware that controls and simulates the inputs. The 'experiment server' is the server that contains the server-side logic, which may include laboratory-specific code, the user interface, the RLMS (Remote Laboratory Management System), etc. The browser is the device through which the end-user interacts, rendering the UI and allowing them to access the laboratory. The following figure does not show the webcam as it is not obligatory in all remote experiments, but it should be regarded as part of the remotisation hardware.

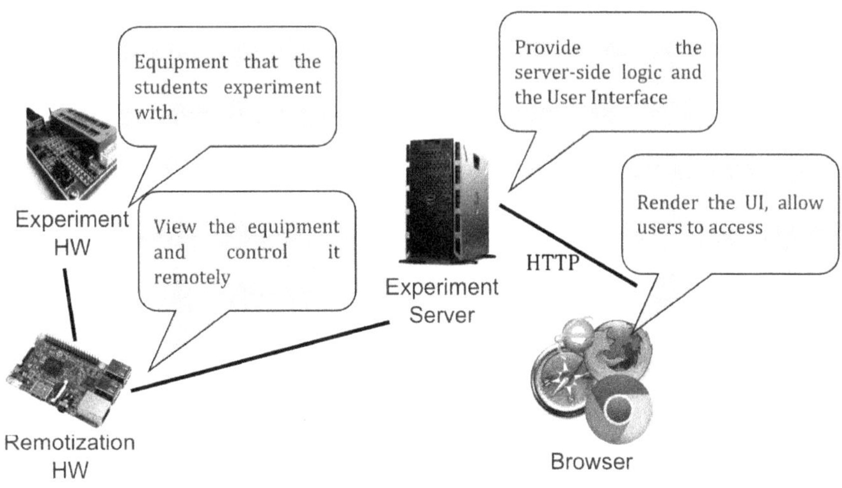

Typical components of a web-based remote laboratory.

The rest of this Part 3 is divided into three parts: general requirements and recommendations for the designer (Chapter 5), architecture chosen for the design of remote laboratories and examples of design (Chapter 6).

Chapter 5

Requirements and Recommendations for Remote Laboratory Design: Overview and Best Practices

This chapter provides an overview of the remote lab development process, along with a few best practices and bad practices for each sub-process. This is intended to serve as an overview for deciding what is needed and what to focus on at each stage and to help teams with limited remote labs experience in planning new remote lab projects. Many of the concepts and requirements in this chapter have already been introduced in Chapter 2, and it makes sense to return to them to relate them to technological aspects of the design. As an example of objectives, consider the approach adopted by the Texas Southern University (TSU) designers, which can be seen in Fig. 5.1.

Figure 5.2 summarises some of the key aspects and processes in a remote laboratory development project. Conceptually, this could be seen as a set of processes that takes place in order from left to right, though in practice this tends to be more complex. 'Pedagogical design' involves the first steps before starting the creation of a lab: choosing what the laboratory will be for and what kind of interaction users will require. The 'HW design and development' tasks and the 'software design and development' tasks are

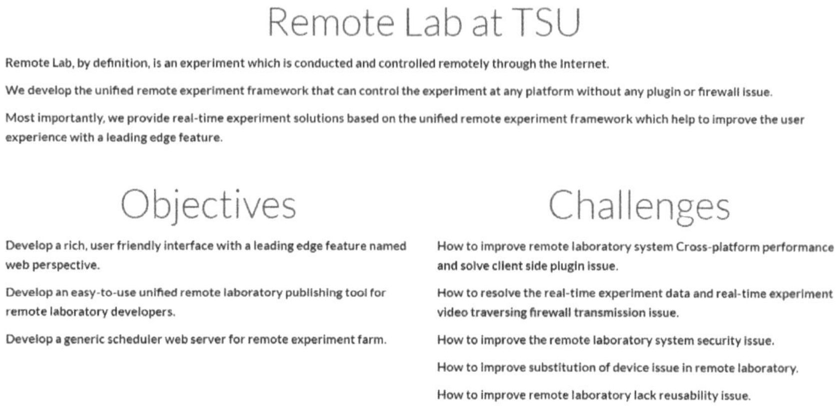

Figure 5.1. Objectives of the Texas Southern University remote laboratory.
Source: http://vr-lab.tsu.edu/index.php/Remote_Lab

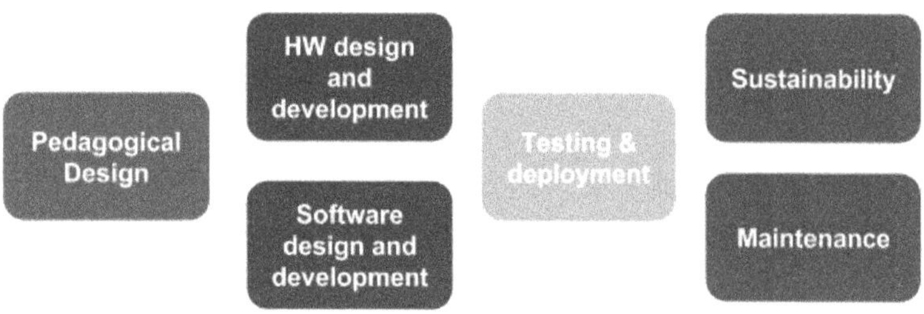

Figure 5.2. Diagram summarising some of the key aspects and processes in a remote lab development project.

self-explanatory. The 'testing and deployment' tasks involve ensuring that the remote laboratory meets the requirements, including the pedagogical ones, and deploying it appropriately. The last two aspects, 'Sustainability' and 'Maintenance', are critical for any production-level remote laboratory and are probably the ones that are most often neglected. 'Maintenance' involves those tasks intended to keep the laboratory in working order, which will include, among other tasks, hardware repairs and software updates. Proper maintenance requires the availability of certain resources, including financial ones and work hours. 'Sustainability' refers mostly to the plan and scheme by means of which these resources are meant to be obtained in the future.

5.1 Pedagogical Design

Before starting the actual development of a remote laboratory, the pedagogical goals it should meet need to be defined. Before moving to the actual development stage, some best practices would therefore be the following:

- Know and define the laboratory's intended audience.
- Know and define what concepts the laboratory should teach and include.
- Try to verify with the intended audience whether the proposed laboratory will meet these goals.
- Design broadly how users will interact with the laboratory and what actions they will be able to conduct.

5.2 HW Design and Development

Some general recommendations for this stage would be the following:

- **Use an architecture such as the one proposed in Fig. 6.1.** Minimize the hardware/software coupling.
- **Keep costs down unless there are good reasons not to.** Even if there are significant resources during the development stage of the lab, the resources available for maintenance tend to be lower. Expensive equipment might need to be replaced in the future. Note that a good reason, however, would be precisely to choose more expensive but more reliable hardware.
- **Avoid over-engineering.** As a rule of thumb, try to keep the hardware as simple as possible (while meeting the design goals) and avoid adding features that are not needed. Generally, those features will tend to increase equipment costs, increase the complexity of the project and maintenance costs, and decrease reliability.

- Ensure that a hardware design and development expert conducts or at least reviews and corrects the work.

- **Design the hardware to be reliable.** This is particularly important for remote labs because they are expected to work 24/7 and simultaneously. Users have more issues trying to get the problem fixed than they would if they were physically present.

- **Consider adding auto-diagnostic hardware.** It is important to be able to know, ideally in an automatic way, whether the laboratory is working or not. No matter how well designed and developed a remote laboratory is, it will almost invariably breakdown eventually. The next best thing to preventing the latter is to at least detect when it happens.

- **Design the hardware to make maintenance easy.** Maintenance is critical. There is a tendency towards limited resources during the maintenance stage, and significant hardware changes are harder to make once the development itself is over.

- **Consider adding admin-only remote-reset and turn-off capabilities.** Preventing hardware failure would theoretically be a better approach, but in practice this is hard to achieve. In practice, a restart can often solve certain immediate issues. Being able to remotely turn off the equipment can also be useful.

- **Plan for replication.** Another reason to keep costs down is that if the laboratory is successful, you might want to replicate it. Through replication, the laboratory will be able to serve more simultaneous users, and you will be able to provide fault-tolerance, because one instance can be turned off if it breaks, while the other instances handle the load.

- **Design the lab for easy deployment.** Deployment means that a remote laboratory can be deployed in another country or institution. This will be important, again, if maintenance is required and if the lab needs to be replicated.

5.3 Software Design and Development

Some general recommendations for this stage would be the following:

- **Dedicate a software expert.** Although it can happen, generally the person that develops the hardware will not be a software expert as well. The software part of a remote laboratory is far from trivial (in many remote laboratories, the amount of time required to handle the software side properly is in fact significantly higher). Neglecting this tends to result in unusable and unreliable laboratories, unprofessional user experiences, poor scalability, unjustified deployment restrictions, bad results in general, and even security risks for which the institution could be liable.
- **Use an architecture such as the one proposed in Fig. 6.1.** Minimize the hardware/software coupling.
- **Ensure that you can test the components separately.** Ideally, make sure you develop and run automated tests (and, in general, apply general software development best practices).
- **Ensure that you can develop and design the laboratory without requiring access to the hardware.** This can often be done by simply hard-coding the REST service responses if you use an architecture such as the proposed one, or through simulating hardware responses.
- **Create for the web.** Unless there are very strong reasons not to, the UI of labs should be web-based and should not require plugins.
- **Create for mobile.** Ensure your design is ideally responsive and mobile-compatible. Nowadays, more people access the Internet through mobile devices than through laptops or PCs, and that trend is expected to continue. Even if your intended audience is PC-only at the moment, there is a chance that this will change in the future.

- **Avoid end-user proprietary plugins.** The UI should not depend on plugins such as Flash or Java Applets. Besides outdated, those plugins are very often not mobile-compatible, sometimes they require admin-privileges to install, and they are almost inherently security risks.

- **Ensure security, from an equipment perspective.** Users, willingly or unwillingly, might try to conduct actions that risk breaking the equipment. Remote laboratories give control over hardware, which means they involve particular security risks. Proper software development is therefore important to avoid the latter. It is advisable to ensure that the hardware is safe from multiple layers, especially from those closest to the hardware. Security should never rely on the client UI.

- **Ensure security, from an end-user perspective.** As is the case with any software that is intended for end-users, efforts must be made to ensure that they are not exposed to security risks. This is of particular importance for remote labs, because users are often students from educational institutions, and sometimes even minors. The institution that hosts the laboratories can often be, in fact, legally responsible for security and therefore for issues that might arise. This is one of the reasons why, for example, exposing students to security threats such as proprietary browser plugins is unadvisable.

- **Design and plan for deployment.** The deployment of a remote laboratory is subject to several particular constraints that may not be immediately obvious to new or even experienced remote lab developers. Some institutional networks, for instance, often restrict inbound or even outbound ports. In practice, for example, if a remote laboratory relies on ports other than 80 and 443, it will be unusable in many institutions. Likewise, the IT services of both provider and consumer institutions will impose their own requirements. As a rule of thumb, for example, do not expect end-users to be able to install any software on their machines.

5.4 Testing and Deployment

Once the hardware and software of a remote laboratory have been developed, laboratory creators might be tempted to consider the laboratory as 'done'.

However, the testing and deployment stages are critical for a remote laboratory that is intended to provide real educational value to students for some time to come.

Figure 5.2 is meant to provide a high-level view of the stages, so testing and deployment are placed together in a single isolated box. In practice, it is of note that while the initial deployment and much testing will take place just after the development stages, testing will also be an important part of maintenance.

The final goal of testing is, broadly, to ensure that the laboratory meets its design requirements: particularly, that it can provide the learning experience that it was designed for. For this reason, at this stage, it is highly advisable to have a relatively thorough specification of the activities that should be able to be conducted with the laboratory and the kind of results that should be obtained. These can then be used to formulate certain tests that can be used to verify whether the laboratory satisfies those requirements or not. These tests should be particularly thorough just after laboratory development.

Once the laboratory has been confirmed to work as expected and maybe tested for a period of time, it should be deployed in its production environment. Here, we include some best practices. It is not necessary to follow them all, or to follow any of them at all, though it is advisable to consider them and to understand their intended purpose.

- Certain software components can often be deployed in the cloud or in a commercial or institutional data centre. This is often preferable since those centres often provide higher reliability, particularly regarding network and power supply.

- The hardware of the remote laboratory should ideally be accessible only to those responsible for maintaining it. A convenient way of achieving this is to have the production laboratories in a purpose-specific environment in a locked room. The goal is to avoid accidental interactions with the equipment that can disable the laboratory, which are more common in a shared environment than one would probably guess. Examples of those interactions are, e.g., people touching the camera and making it point somewhere else (disabling the laboratory), putting some object in the field of view of the camera (disabling the laboratory), turning off the lights of the room or the laboratory (disabling the laboratory), putting some object that blocks the expected behaviour of the laboratory, etc.

- Fix the components of a production laboratory so that they cannot be accidentally moved from their intended positions. This is particularly important for the cameras, but also for other components. Some alternatives to accomplish this are to place the laboratory in a box (if small enough), to fix the camera or particular parts with screws to a main body, or to fix them to a vertical stand.

- Ensure nonetheless that the production environment is accessible to the laboratory maintainers, in case the laboratory needs to be repaired.

As mentioned before, after successful deployment, testing is not really finished. While it does not need to be as thorough, testing will remain important throughout the lifetime of the laboratory. Tests should determine whether the lab still works as expected, so that it can be disabled and repaired if it does not. Some key aspects to consider are the following:

- It is preferable for tests to be automated. That way they can be run periodically with less effort, or they can be run when the laboratory is suspected to be failing.

- The best approach when a laboratory fails is often to turn it off so that users can be warned that it is under maintenance.

- The worst possible failure from a user perspective are not failures that prevent use of the laboratory, but failures that run undetected and that result in the lab being accessible, but providing incorrect results. Such failures mean that students will waste their time conducting an experiment that cannot succeed, at best. At worst, they will believe the laboratory to be working as expected, be misled by incorrect data, and draw wrong conclusions. This produces an extremely poor user experience that will strongly discourage students and their professors from trusting or trying to use the laboratory ever again.

- Because of the above, it is particularly advisable to be able to detect automatically failures that can result in the experience above, so that the laboratory can be disabled and repaired.

It is sometimes particularly useful to automate tests for those components that fail at times. That way, not only will the failure be detected easily, but maintainers will know straightaway what they have to repair, e.g., a web camera.

5.5 Maintenance and Sustainability

A problem facing some institutions is the fact that they tie the development of their remote laboratories to specific projects: either funded ones (such as EU projects) or internal ones (end-of-degree projects, etc.). Even if the remote laboratory reaches production-level before the project ends (which is something not to be taken for granted, especially for end-of-degree projects), when the project is over there are often no actual resources for maintenance and no viable sustainability plan. These laboratories tend to be of little use, especially for non-internal consumption; they, thus, tend to be short-lived.

Maintaining a laboratory is critical, and it is a fact that maintaining a laboratory requires resources (in many cases, significant resources), even if the laboratory is designed to be particularly stable.

Depending on the circumstances of each remote laboratory developer and institutions, different strategies might be employed to attempt to guarantee sustainability. Some ways to try to achieve this sustainability will be listed here. Some may or may not be possible or likely to succeed, depending on the circumstances.

- Guarantee maintenance funding from the lab developer institution, often a university. This is sometimes difficult because there are often funding limitations, and because university administrations seem to have significantly fewer issues with 'one-off' expenses than with more gradual but prolonged investments.

- Ensure that the laboratory is used internally by as many teachers and students as possible and that its benefits are clearly documented. This will often act as an encouragement for maintenance funding.

- Ensure that the remote labs group is recognised as a research group (or linked to one) within your institution and that it is funded as such.
- Contact third parties who might help provide a sustainability scheme, such as LabsLand.

LabsLand (https://labsland.com) is a company and a network of remote labs. LabsLand offers a large set of remote labs that are created in collaboration with different institutions or provided directly by the latter. One of the original reasons for founding LabsLand was, precisely, the difficulties that academic research groups and institutions have in fulfilling their ultimate objective: achieving reliable, trustworthy and sustainable labs. This is described in more detail in the following subsection.

5.6 LabsLand and Sustainability

LabsLand (https://labsland.com) is a start-up and a spin-off from the WebLab-Deusto research team at the University of Deusto, set up in in 2015. It is also a network of remote labs from providers all over the world, which focuses on providing production-quality labs that are reliable and scalable. LabsLand offers more than 30 experiments that are federated in different countries and institutions.

The LabsLand business model involves being a marketplace with both lab consumers and lab providers. Consumers simply pay a subscription fee to use the labs in the network. Providers are those who collaborate with us by making their labs available on the network and benefit from a compensation mechanism.

LabsLand currently has partners in over 14 countries in various continents. Depending on each case, the benefits to the partners vary: monetary compensation, visibility or free access to other laboratories.

The remote experiments that are part of the LabsLand network must meet two relatively obvious but very significant criteria. First, they must be reliable. Users and teachers expect the labs they are going to use to work. Historically, this has been one of the reasons why virtual labs have generally been more popular than remote ones. LabsLand remote labs, however, are generally just as reliable as a virtual lab, as, to begin with, the LabsLand real-time laboratories are always in multiple universities at the same time.

Second, the labs must (generally) scale to a high number of users. This is often handled on a case-by-case basis. If the laboratory is interesting enough, LabsLand often helps the partner institution achieves that goal through replication.

In general, adding a laboratory to LabsLand involves no drawbacks for any partner, and only has a considerable number of benefits.

5.7 Goals and Requirements

The main design goals of the remote laboratory design from this book's perspective are the following:

- **Decoupling.** The remote laboratory will often have to be developed by at least two different experts: one hardware-focused and one software-focused. The architecture should split the remote laboratory into several micro-components. The components can be developed independently by the most appropriate expert, tested independently and then integrated with relative ease.

- **Reliability.** A production-level remote laboratory needs to be reliable. One of the key obstacles to mainstream remote lab adoption, and the reason for the popularity of virtual labs, is that reliable remote labs are relatively rare. Remote lab owners often find that 24/7 maintenance of labs online is a considerable challenge, especially once the projects that often led to their creation are over. A lack of reliability can stem from deployment issues, hardware failures, software issues, difficult maintenance, network issues, etc. The architecture should therefore be oriented towards minimizing those problems.

- **Scalability.** This refers to the capacity simultaneously to serve a high number of users. This is particularly challenging for a remote lab, because, often, the underlying hardware can only be used by a single student at a time. The architecture that is going to be described should therefore be oriented not only towards achieving scalability in its software but also towards making replication easier in order to be able truly to scale the number of users. Replicability is also a way of promoting reliability.

- **Security.** The deployment and use of a remote laboratory should not compromise the security of the student using it or of the institution where it is located.

Chapter 6

Remote Laboratory Architecture

Figure 6.1 provides an overview of the proposed architecture, the key aspect of which is that it comprises several different components that are designed to be decoupled and developed and tested independently. The 'remote lab hardware' includes both the experiment hardware and the remotisation hardware. It is very laboratory-specific and is controlled by the device server, through which it communicates directly in a laboratory-specific way. For example, in the pendulum experiment, the device server is responsible for giving the order to run the motor that raises the pendulum, which is the hardware part, or in VISIR is responsible for indicating which relays should be opened and closed, which are the hardware part.

The 'Laboratory Server' is the laboratory-specific component that handles the higher-level laboratory logic. It is decoupled from the hardware: it does not communicate with any specific hardware directly. Instead, it controls the hardware indirectly through a REST API that is agreed with the 'Device Server'. This component relies on the WebLabLib Open Source library, which provides, among other features, the capacity

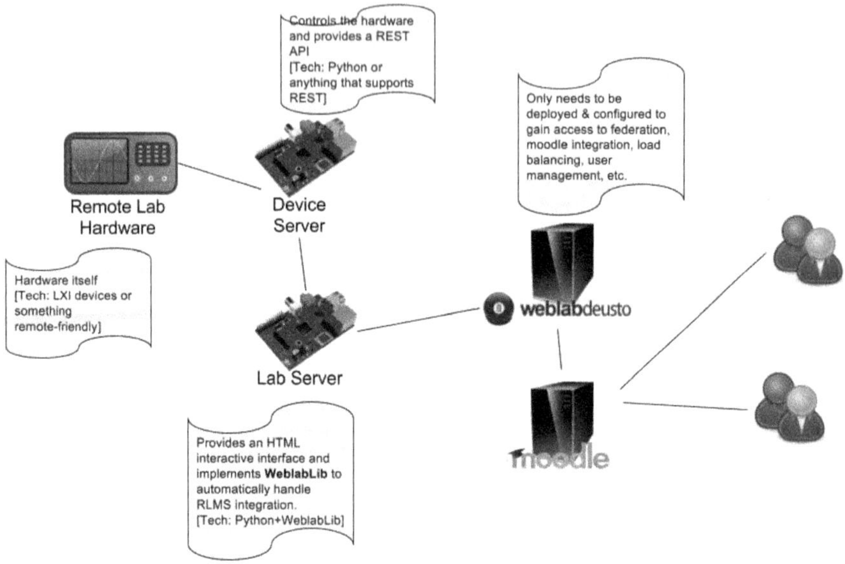

Figure 6.1. WebLab-Deusto architecture overview.

for automatic integration within a RLMS system. The RLMS system, depicted in Fig. 6.1 as 'WebLab-Deusto' (https://weblab.deusto.es/), can even be deployed in the cloud and provides lab-generic common features such as authentication, user management, Learning Analytics, LMS integration, etc. Users access the lab either directly or through an LMS (e.g., Moodle), in which the laboratory can be automatically integrated by relying on WebLab-Deusto.

6.1 Architecture Layers

The proposed architecture is divided into three different layers, which are shown in Fig. 6.2.

The layers are all designed to be loosely coupled and to ease practical remote laboratory development. The hardware layer contains the experiment and remotisation hardware, and only very low-level software. It communicates with the lab server layer through a simple REST API. The Lab Server layer contains the laboratory-specific logic and software (this includes the UI, the actual server-side laboratory logic, etc) and relies on the WebLabLib framework to provide certain functionality and automatic

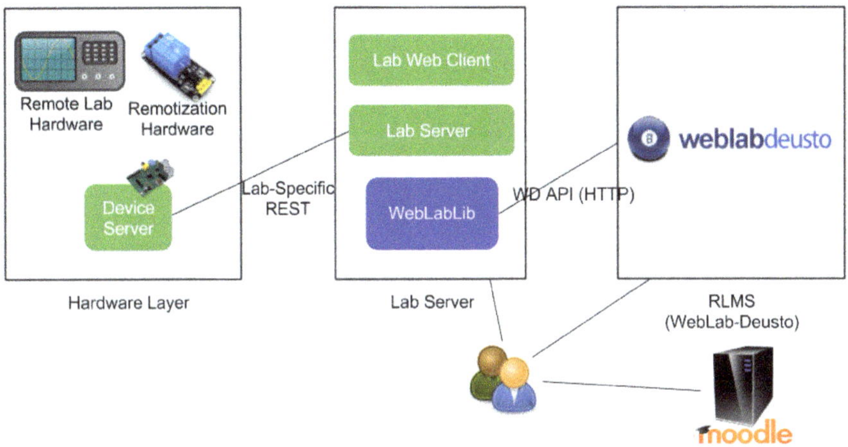

Figure 6.2. Layers of the WebLab-Deusto architecture.

integration with the WebLab-Deusto RLMS. The RLMS layer only needs to be deployed and configured, as it is generic for any laboratory. It does not contain laboratory-specific code. It communicates with the previous layer automatically through WebLabLib (which internally relies on an HTTP-based API) and provides common functionality such as user management and LMS integration.

The following subsections describe each of the layers in more detail as well as the purpose and benefits of organizing the architecture with such a scheme.

6.1.1 *Hardware layer*

Figure 6.3 shows the hardware layer. As previously described, this layer contains three main components. First, the remote lab hardware. This is the subject of the experimentation itself, the actual hardware that students are meant to control and view through the Internet. Second, the remotisation hardware. That is the hardware that is actually needed to allow users to control the equipment hardware remotely. Third, the device server. This is a relatively simplistic low-level software component that is meant to be used as an interface to control the hardware. Generally, the device server will not be computationally expensive and will only need to serve a limited number of simultaneous requests, so an inexpensive device, such as a Raspberry Pi, will often suffice. However, depending on the laboratory, it might need to

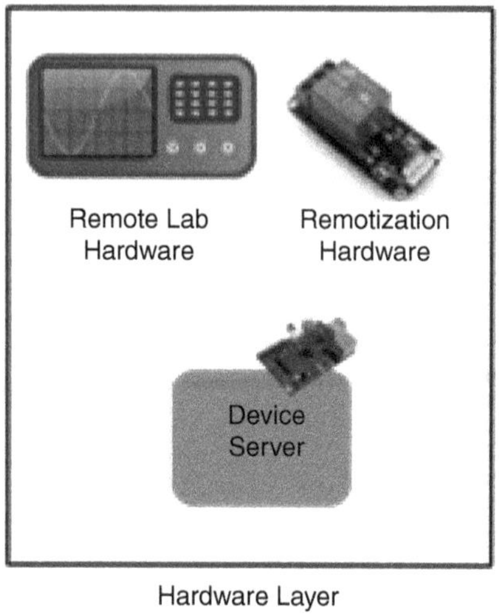

Figure 6.3. The WebLab-Deusto hardware layer.

interface with different hardware, and will therefore require the means to do so. In the case of Raspberry-based laboratories, they often rely on the GPIO pins or on USBs to interface, but this may vary.

One of the key characteristics of this layer is that it is mostly hardware. Properly development of this layer, therefore, will normally require a hardware expert but not a software expert. The device server contains some software, but it is low-level (close to the hardware) and is designed to communicate with the next layer through a simple REST API. This means that the internal device server architecture, from a software perspective, can be extremely simplistic. It simply needs to be able to respond and handle HTTP requests. It does not need to deal with external dependencies or libraries, it can be implemented in any programming language and it does not need to concern itself with scalability, user experience (UX), user interface (UI), or any other goal than to specifically control the hardware remotely. Likewise, because the layer does not really have any external dependencies, it can be tested independently. The developer of this layer can test the layer by simply making the proper HTTP requests. If appropriate, this could even be automated with relative ease. Therefore, if properly tested beforehand, the integration process with the next layer should be particularly straightforward.

6.1.2 *Laboratory server layer*

Figure 6.4 shows the laboratory server layer. This layer is completely decoupled and isolated from the hardware. Generally, however, it will still be deployed close to the actual hardware. That way, if they share a private network with the hardware layer, the lab-specific REST API will not require authentication or session management, and latency can be minimised. There are three key components in this layer, which are all in the same physical machine. This machine can be any kind of server, and often a simple Raspberry will be powerful enough to host them as well. The 'lab web client' is the actual UI through which users will interact. In a modern laboratory, the UI should generally be fully web-based. The 'lab server' is the server that contains the lab-specific logic. It should rely on the WebLabLib framework to provide some basic utilities, it controls the hardware indirectly through simple HTTP requests and the lab-specific REST API (which connects to the Hardware layer) and it is automatically integrated into the WebLab-Deusto RLMS (WebLabLib handles that the latter).

The web client can be developed in almost any web technology. In LabsLand, as of now, most labs are developed using standard JavaScript/HTML/CSS with either jQuery or AngularJS. The laboratory server could, in principle, also be developed using any technology, though Python is highly recommended in order to use WebLabLib. All LabsLand laboratories are developed using Python, Flask and WebLabLib.

Similarly, but in contrast to the hardware layer, the key aspect of this layer is that, in this case, no hardware is directly involved. Therefore, a

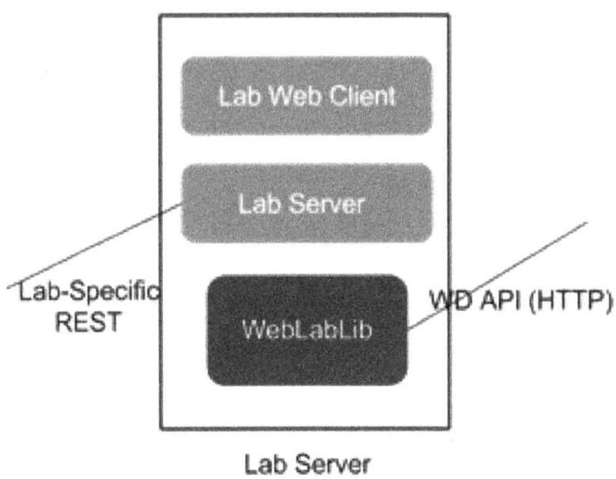

Figure 6.4. The WebLab-Deusto laboratory server layer.

software expert is required. Both layers communicate through the REST API, which should be agreed upon beforehand. From that perspective, this layer can also be tested independently. WebLabLib provides the facilities that are required to simulate user sessions and to develop the laboratory without even needing to install the actual RLMS. The fact that the hardware is accessed only through a REST API makes it relatively straightforward to test this layer independently of the hardware, by simply simulating or hard-coding the REST API responses.

Therefore, the laboratory software developer only needs to worry about the laboratory logic and not the hardware, developing the UI independently, etc. Other than for the REST API design and eventually the integration stage, the hardware and the lab server layer developers can develop their respective layers independently, in parallel if required, without depending on each other.

6.1.3 *RLMS layer*

Figure 6.5 shows the RLMS (Remote Laboratory Management System) layer. An RLMS is a software tool that facilitates the development, maintenance and deployment of remote laboratories. The main purpose of an RLMS is to provide out-of-the-box features that are important and that are common

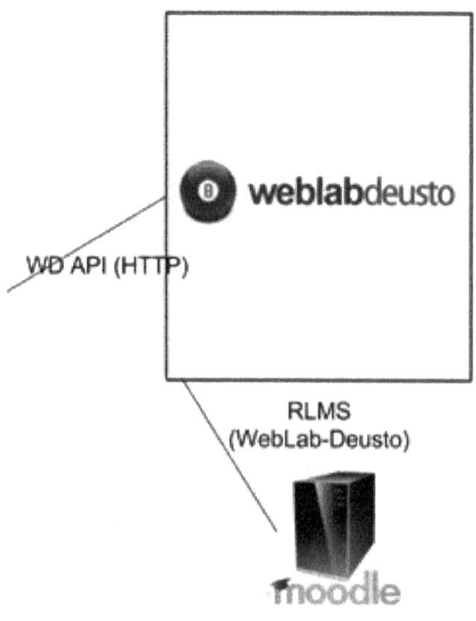

Figure 6.5. The WebLab-Deusto RLMS layer.

to all or most laboratories. These are features such as user management, authentication, scalability, Learning Analytics, federation and sharing of laboratories across different institutions, integration within Learning Management Systems, and the like. The RLMS that we recommend is WebLab-Deusto. By integrating a laboratory with WebLab-Deusto, the laboratory automatically obtains these features without needing to develop, integrate or maintain them.

The key aspect of this layer is, therefore, the fact that, unlike the other two layers, it is mostly laboratory-agnostic. The WebLab-Deusto RLMS server needs to be deployed, and can in fact be deployed in the cloud (this is actually recommended). Communication with the Lab Server layer is handled transparently and automatically through the WebLabLib library. Essentially, a laboratory that is based in WebLabLib is automatically integrable with WebLab-Deusto. Therefore, once this layer is deployed, the laboratory simply needs to be configured, and no particular development is required here.

Figure 6.5 also shows a Moodle LMS. This is depicted as an example, because students will often not access the RLMS directly, but instead access it through an RLMS. This way, teachers can integrate a laboratory within their Moodle course and students can access the resource transparently. It is also worth noting that, although Moodle is given as an example, many more LMS, like Sakai, Google Classroom, etc. are supported.

6.2 Cases Studies of Remote Laboratories

This section particularises the aforementioned architecture for specific remote experiments. This will provide clarification of various aspects of the architecture described.

In these examples, emphasis will not be placed on the RLMS part, since if the design is assumed to be decoupled, then the designer can integrate it within any available RLMS, for instance, and in our case, WebLab-Deusto. In this way, all the LMS characteristics are obtained, without affecting the other part of the experiment. In other words, if the designer installs WebLab-Deusto, they automatically meet all the requirements previously stipulated for Layer 3, and one simply needs to design Layers 1 and 2 using WebLabLib.

Consequently, this section is dedicated to the low level part, to the remotisation of hardware, device server and lab server.

First, some examples are presented (Sections 6.2.1–6.2.5), followed by a description of the technological solutions available (Section 6.2.6). The examples included do not necessarily have to exist or be part of any active remote laboratory. They are examples of types of laboratories based on Chapter 3.

6.2.1 *Sensored remote experiments*

Imagine an aquarium with electronic temperature sensors, dissolved oxygen, turbidity, pH, conductivity, presence of salts, etc. Many of these sensors are equipped with connectivity and publish their results in open fashion on the web. This is the case of the IoT devices or the Pasco brand sensors (https://www.pasco.com/products/sensors). And if this is not so, it is easy to compile the data from the sensors and transfer it to an interface.

In this case, the remote experiment may not require additional services. The student will simply connect to the IPs of the sensors, compile the data and complete their task, which might be to relate the measured signals: is conductivity related to the presence of salt? And the turbidity and the presence of oxygen?

One example is the weather station offered by the ISES remote experiment (http://www.ises.info/index.php/en/laboratory/experiment/meteorological-station-in-prague/) where the student learns to compile and relate meteorological data.

Another example of a different experiment from the former is the planarian experiment (see Section 3.2.1). In this case, the student can only see and measure a planarian's activity in one of various solutions, and does so by means of a web camera. If the latter has an IP, then the architecture is reduced to nothing and the designer only needs to give the student the camera IP address. A good interface will, however, clearly improve user experience.

6.2.2 *Batch experiments*

In this case, and as was seen in Section 2.1.2, the user parametrises an experiment in its inputs, sends them and subsequently receives the results thereof.

A typical example of batch experiments are experiments of characterisation of electronic devices, for example a diode. In the simplest case, the student indicates that the diode should be supplied with a voltage ranging from −1 V to +2 V with an interval of 0.1 V. The experiment will return the current measured for each supply voltage.

In this experiment, one can use power supply units manufactured with LXI or PXI technologies, which are more expensive but allow the user to control their behaviour in transparent fashion via a webpage or an IP. In other words, the designer does not have to address the 'hardware remotisation' task and neither is a device server necessary since both are built into the power source by the manufacturer. Moreover, manufacturers usually follow standard communication protocols and all allow for control and reading via the Internet, which facilitates the design and deployment of the remote experiment. This equipment used to be expensive (thousands of euros), but is now available at a very reasonable price. A piece of advice: do not design something that is already manufactured by a recognised company.

In this case, the fundamental work consists in designing an interface that is agreeable to users and facilitates their work.

6.2.3 *Interactive and controlled experiments*

The more complex the service something offers, the more complex the architecture. The hardware part should allow the student to upload a program, interact with the platform and observe the latter's behaviour via webcam.

In order to teach students how to program an Arduino board, they should know how to write programs, compile them, upload them in the Arduino, activate the buttons and switches, and see the results. To remotise the Arduino, it is necessary to address all the aforementioned tasks.

We can begin by suggesting students download the Arduino environment, and thus write and compile their code on their own computer. When the .bin file or similar is obtained, the remote experiment task begins.

The device server should contain the Arduino loader so as to upload in its memory the student's .bin file. This may be no trivial task for other programmable devices.

This .bin file will have been uploaded using the interface designed in the lab server. This interface will invite the student to click on buttons and switches, which will send the corresponding command to the device server in order to activate the corresponding inputs and observe their effect on LEDs or motors via a webcam that will be connected to the interface of the lab server, or may transfer its content directly to an IP.

The previous example is simplified for a robot controlled by an Arduino, since in this case the student should not interact with the robot, but simply upload a .bin file and observe the robot's evolution via webcam.

In this case, there is no remotisation hardware (except the webcam), as all the input/output signals are controlled by the robot microcontroller (Arduino, ARM, etc.). However, if for example the task were to program a robot to escape from a maze, then an additional control would be required to take the robot to the maze entrance for each user; this control would be remotisation hardware.

6.2.4 *Electromechanical experiments*

A special case, different from the previous examples, is when there are electromechanical elements, whether for batch or interactive remote experiments.

For instance, in Archimedes experiment, it is necessary to raise and lower an object inside a test tube. In a robot, one has to move robots. In the inclined plane, it is necessary carefully to position a motor before launching the ball.

In all these cases, the remotisation hardware and the device server are extremely important. If, for example, the Archimedes lab server interface says the ball should be lowered, then this command for the motor becomes a PWM signal generated by a programmable device (Raspberry/Arduino/ARM/FPGA...) that has been programmed by the designer. This motor will have a sensor to detect the ball's position, and therefore the designer will have to be careful when programming its control.

This remote experiment is clearly very dependent on the control programmer of the electromechanical device. Both in normal mode and in test or failure mode, the programmer should be experienced.

This situation did not arise, for example, with the LXI/PXI power source, as in this case the command is executed within the source via a service created and tested in the company, which renders it more reliable.

Many remote experiments are not addressed because their electromechanical complexity is very high and because they also involve a number of risks. Chemistry experiments are a good example, as there are substances that need to be poured, stirred, illuminated, etc. These activities are performed manually or by means of electronic equipment not yet connected to the Internet in order to be operated.

6.2.5 *Time-deferred or ultraconcurrent experiments*

Section 2.4 referred to these experiments. They are usually batch or sensored, and cannot be interactive or programmable. However, they are highly recommendable when the experiment is electromechanical or should be multiuser.

This approach was pioneered by Kirstein (1997) in the Free University of Berlin (Germany) with the ISE (Interactive Screen Experiments) concept. An ISE consists of a filmed experiment with the data provided by the experience. This concept was later developed and improved by the Stanford Cyberlab (http://secondilabs.appspot.com/) (Hesselink *et al.*, 2003). In ISE, the experiment was filmed, and, by using multimedia, the data and some kind of interaction with the user were added to the remote experiment (see Fig. 6.6). Subsequently, the user session was recorded: the interface, the webcam, the data, etc.

Using this technology, the experiment and its interface are recorded a considerable number of times for each scenario and evolution of the experiment. So when students access the experiment, they will see a recorded sequence in which they will be unable to distinguish what it is. User experience is complete and in no way different from what is 'normal'.

The question is how is an ultraconcurrent experiment designed? The answer is far from simple, and though it may be of interest to many readers, it is completely beyond the scope of this book. However, and given this interest, LabsLand offers a tutorial and support for the design of ultraconcurrent experiments, since, despite their complex design, the latter

Figure 6.6. Production of an ISE experiment.

Source: https://didaktik.physik.fu-berlin.de/projekte/ibe/entwicklung/index.html.

offer significant advantages. Torre *et al.* (2020) also describe the design of ultraconcurrent experiments, called digitised experiments.

6.2.6 *Examples*

The following figures (Figs. 6.7–6.10) reiterate some of the previous examples with a view to clarifying them. In all of them, WebLab-Deusto will be employed as RLMS.

Figure 6.7 shows the architecture of an aquarium as sensored experiment where there is no remotisation hardware:

- The student accesses the experiment via the LMS of their educational centre thanks to the RLMS WebLab-Deusto. WebLab-Deusto records all the student's activity.
- The sensors deliver the temperature value and the % of oxygen in the water to the device server, which can be a Raspberry.
- If the sensors deliver an analogical signal, then the device server uses an A/D converter to convert it into binary information.
- If the sensors deliver the signals directly in binary form, then the device server has less workload.

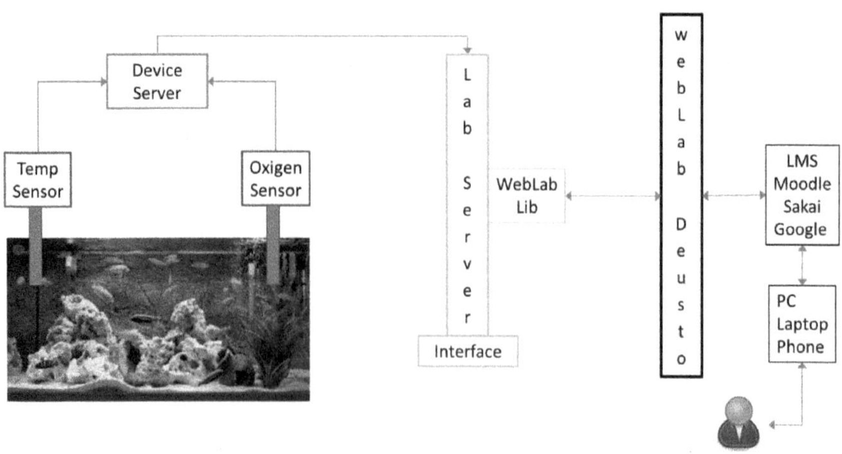

Figure 6.7. WebLab-Deusto architecture for a sensored remote experiment.

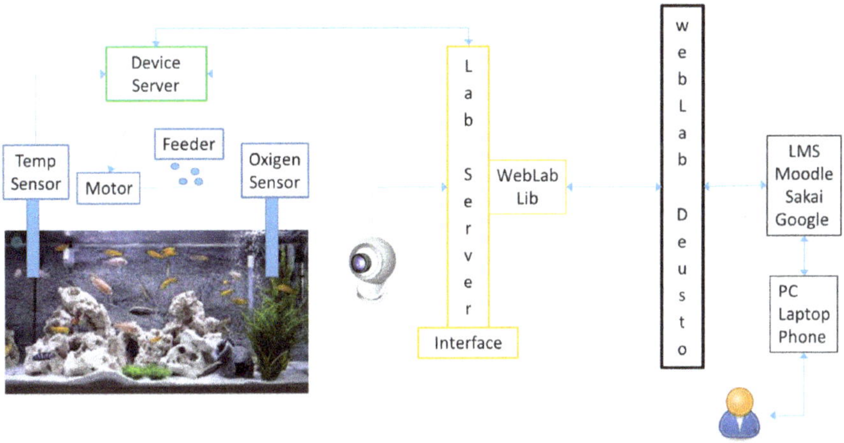

Figure 6.8. WebLab-Deusto architecture for an interactive remote experiment.

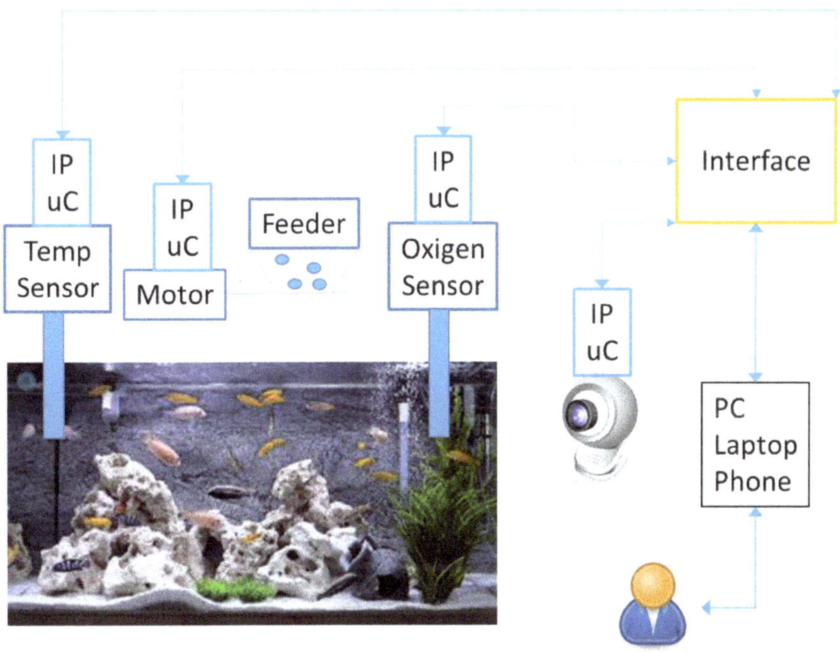

Figure 6.9. WebLab-Deusto IoT architecture for an interactive remote experiment.

The device server sends this information to the lab server. In IoT technology, the device server is called IoT node and manages various sensors.

The lab server integrates that information in the user interface.

The user writes down the necessary data and completes the experiment according to the instructor's instructions.

If we now add a webcam to the aquarium and the user can also feed the fish, the figure changes and the device server's actions increase as there is now remotisation hardware. If the student presses the button to feed the fish, the device server should control the motor that opens the food deposit via a PWM software algorithm, for example. In Fig. 6.8, the sensored remote experiment becomes interactive.

In this case, if overfeeding of the fish should not be permitted, then the lab server should control this aspect via experiment logic.

In Fig. 6.9, the webcam, the sensors and the motor have microcontrollers associated and an IP (as occurs with IoT). The designer also wishes to lighten the architecture by eliminating the RLMS. In this case, the following happen:

- The device server is not necessary since its functions — centralised in the previous case in an IoT node — are distributed in each sensor and actuator.
- The lab server only has the interface, and this may be very simple.

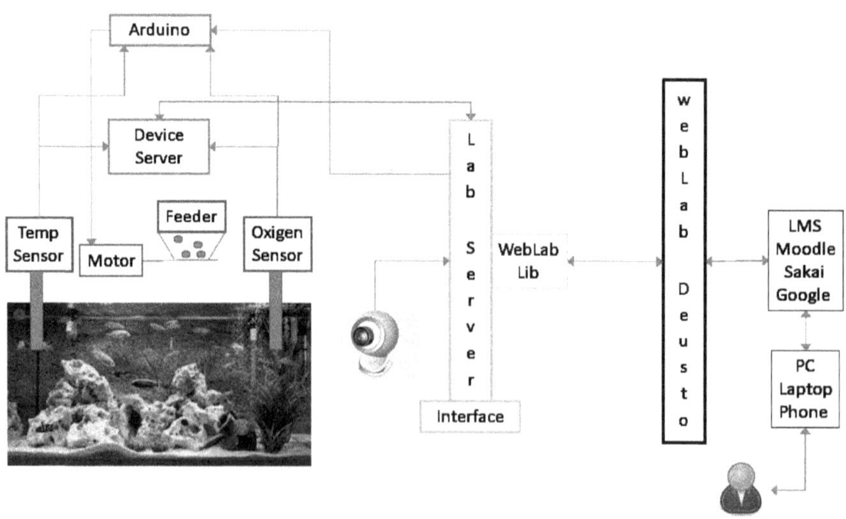

Figure 6.10. WebLab-Deusto architecture for a controlled remote experiment.

- Communication between the interface and the sensors/actuators will be performed via a LAN (Ethernet or WiFi).

This approach gives rise to a plug&play as described by (García-Zubía *et al.*, 2008). The remote experiment is connected to a LAN, which returns an IP, which will be the one the user has to access. This approach is based on the use of web sensors offered by web servers. A good example is a Raspberry, but it could also be a PIC or an Arduino with communications, even if they are not, strictly speaking, web servers. The greater the designer's experience (which should be considerable) and the more standard the equipment used, the simpler the design.

A technological disadvantage of this light solution is that the remote experiment will almost certainly need a public IP and require some ports to be opened, which will not be acceptable to the IT services in the place of deployment.

In Fig. 6.10, an Arduino is added to control the aquarium, and therefore the remote experiment changes from interactive to controllable. In this case, the student should write a Scratch (or Blockly, Python, etc.) program that controls the fish feeding in accordance with their needs and the presence of oxygen in the aquarium.

- The interface becomes complicated, as it should offer the student a Scratch programming environment. There will almost certainly be two webpages: one for programming and another to monitor the aquarium's evolution.

- The device server should assume the task of programming the Arduino.

- The lab server in its experiment logic may need to evaluate whether or not the program involves any risk for the fish and the aquarium.

- The lab server should not allow more than one student to control the aquarium. It should manage the queue if there are students waiting, and assign a fixed time for each session.

- The feeder motor is now controlled by the Arduino.

- The sensor signals should go to the Arduino to execute the program, and also to the device server so it can send them to the interface that monitors the aquarium.
- WebLabLib facilitates communications between the user, the interface and the hardware devices controlled by the Arduino.

In Fig. 6.11, the designer opts to use LabVIEW and NI (National Instruments) data acquisition boards. In this case,

- The designer should know how to programme in LabVIEW.
- The designer should buy the equipment and the software licence to develop the equipment.
- The designer should buy the additional NI software to publish the aquarium control interface on the Internet.
- If the LabVIEW interface is going to be used, then the lab server is not required.

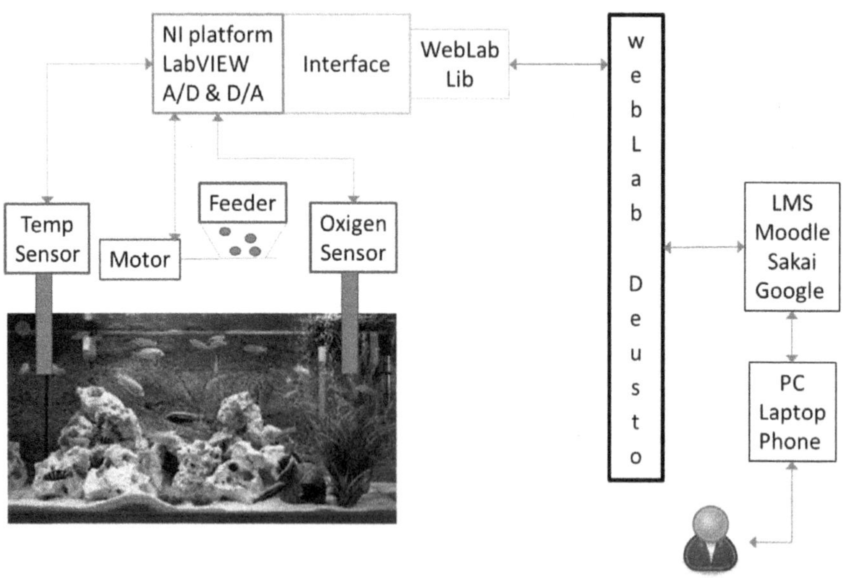

Figure 6.11. WebLab-Deusto LabVIEW-based architecture for a controlled remote experiment.

- The student should know how to programme in LabVIEW if they are going to write the control algorithm. If the design is going to include a supervisor responsible for ensuring that the aquarium and fish are not in danger, then the design becomes considerably more complicated.
- Bearing in mind all the above, the architecture becomes much simpler.
- WebLab-Deusto makes it possible to integrate experiments developed in LabVIEW, and thus access all the advantages of this RLMS using WebLabLib.
- In this case the, device server and the lab server are on the LabVIEW control panel.

If the experiment were batch, the previous figures would not change much. The lab server would simply have to be responsible for enqueueing the experiments to be launched and collecting and publishing the results on the client's interface. In other words, the experiment and the corresponding interface would not be interactive, but more static and, in principle, easier to control.

6.2.7 *Communications and protocols*

In the above designs, the arrows indicate data flow via communication canals and protocols. This is an aspect that should be addressed with particular care from the very beginning as it will condition the efficiency and universality of the remote laboratory. There are three levels of communication: between the hardware and the device server (HW-DS); between the device server and the lab server (DS-DL) and between the lab server and the RLMS.

Communication between the device server and the sensors and actuators (hardware) can follow any protocol, as it does not directly affect the rest of the design. The communication is established via typical hardware protocols like UART, GPIO, ADC, I^2C, SPI, etc. In this field, electronic engineers are very much at home and are usually responsible for developing the hardware part of the remote experiment.

Communication between the RLMS and the lab server (if there is an RLMS) is of high quality and typically Ethernet/IP. This communication is usually or should be in the hands of a software engineer, and it is facilitated by WebLabLib.

The most delicate communication is that produced between the device server and the lab server, especially with its interface: when the user clicks on a button on the interface, how does this command reach the device server in order to be executed? How is the value of a LED changed on the interface? If the designer is an electronic engineer, they will probably opt for protocols like CAN, MODBUS, PROFINET or socket.h library with Ethernet, clearly orientated towards the hardware. But if the designer is a software engineer, then they will probably choose Internet Protocol (IP), Ethernet and web services. A current generic remote laboratory should opt for IP communication, while specific remote experiments may opt for other communication protocols.

Before transmitting the data, to the interface (LED) or from the interface (button), one needs to be in possession of it. From the computer program perspective, to manage hardware information there are the following two basic options:

- **By events.** Every time a specific event occurs, an action is executed, which may be to send or to receive data. The action is only executed when the data changes, in other words, the communication is asynchronous.

- **By polling.** The program continually asks about a datum, whether it has changed or not. Polling is characterised by the time that lapses between one request and the following request, and may be fixed, so communication is synchronous.

With a web services and IP approach, communication between the device server and the lab server may be asynchronous or synchronous.

- **Asynchronous.** Only when an event takes place on the interface does the communication channel open. Control is via http commands like GET and POST, which configure a REST API.

- **Synchronous.** The communication channel is always open, so the interface is regularly modified via the use of web sockets.

One is not better than the other, and each has its advantages. Historically, the use of web services and asynchronous communication with languages like AJAX was a landmark in remote experimentation, while web sockets are an increasingly popular modern alternative.

Before continuing, it makes sense to briefly describe some of the characteristics influenced by the communication channel.

- **Interface performance:** This indicates whether the remote experiment interface is agile and does not freeze or slow down. Performance is directly dependent upon the latency of the signals and the throughput of the communication channel.
- **Latency:** This is measured in milliseconds and indicates the time lapse between an action taking place and its effects being visible. Latency should be a low value because, if there is a long time between the action and its effect, the user's experience will be negative. Latency depends on the throughput of the communication channel and management of the interface.
- **Throughput:** This is measured in megabytes per second and indicates the information transfer capacity possessed by (reality) or required of a communication channel (design requirement). It is best for the value to be high and the requirement low. A low latency requires a high throughput.
- **Reliability:** This is a subjective value that measures the remote experiment's dependency on the client's and the server's network quality. This value is directly dependent on the throughput: if the requirement is high and the real availability is low, reliability is low.
- **Complexity:** This is a subjective value which indicates whether to use a particular type of communication. The designer needs to be experienced and the device needs to be complex and with abundant resources. It is best for complexity to be low.

- **Security and accessibility:** This is a less subjective value than the previous one and indicates whether a type of communication jeopardises IT security or compromises the accessibility of the remote experiment from a web browser, Smartphone, etc.

Let us consider both strategies for a very simple case of an interface: a button (output) and a LED (input).

- **Button:** This is an http of an interface that generates an event when pressed. Control can be in the following ways:
    - **Asynchronous:** Only when the button is pressed does an event occur. The lab server opens the channel, sends the action (button=1) and closes the channel. This is a simple (easy to programme), light and efficient (occupies little of the communication channel) communication.
    - **Synchronous:** The lab server always has the communication channel open with the device server via web socket, and when the button is pressed its value is sent. This is a communication that requires the channel to always be open (not very efficient) and the device server to be powerful (currently the norm). Furthermore, the designer should know how to use and configure web sockets.
    - **Comparative:** The latency is similar, as is the throughput required. The use of web sockets requires a device with more resources and an experienced designer. If the network crashes, reliability is more compromised in the web sockets solution, as the ports should be permanently open. In general, both solutions are similar and it is the designer who makes a choice.
- **LED:** This is an interface, illuminated or not, in parallel with a real LED.
    - **Asynchronous:** As the lab server does not know when an LED change event occurs, the designer should programme data polling between the lab server and the hardware; in other words, open the communication channel every 100 ms

(or another value), read the LED value and send it to the lab server to be published on the interface. This is done even if the value has not changed, and a typical error is to reduce the polling time to reduce the latency, as this will increase the required throughput.

- **Synchronous:** As before, it is necessary to conduct data polling, but this time between the device server and the LED (the hardware), so only when the device server detects a change, does it send the new value via the web socket.

- **Comparative:** The asynchronous solution is worse in this case since if the polling time is reduced, there are many requests and therefore the final latency (because the interface will collapse) and the throughput will increase, significantly reducing interface performance and user satisfaction. If, on the contrary, polling time rises, then it is latency that is directly increased. This solution depends very much on the quality of the network, which does not have to be high. The web sockets solution tends to prevail in these cases.

Figure 6.12 clearly shows the location of the polling mechanism. In REST API, polling affects the lab server and therefore the entire performance of the remote experiment, while in WebSocket polling only affects the device server and therefore has less effect upon performance.

Both strategies can be used in the same remote experiment and within the same remote laboratory.

Another question is the security and universality of the remote laboratory. A first step consists in ensuring that communications between the hardware and the lab server via the device server take place in the same LAN, so management of ports and firewalls is safer and simpler. The best way of achieving this is by placing all these experiments together in the same physical space. For this, the following two options exist:

- **Light implementation.** Integrate the device server and the lab server in the same device, for example, a Raspberry. This is possible because the Raspberry is increasingly powerful and because WebLabLib can be adapted to this situation, reducing the services provided by WebLab-Deusto.

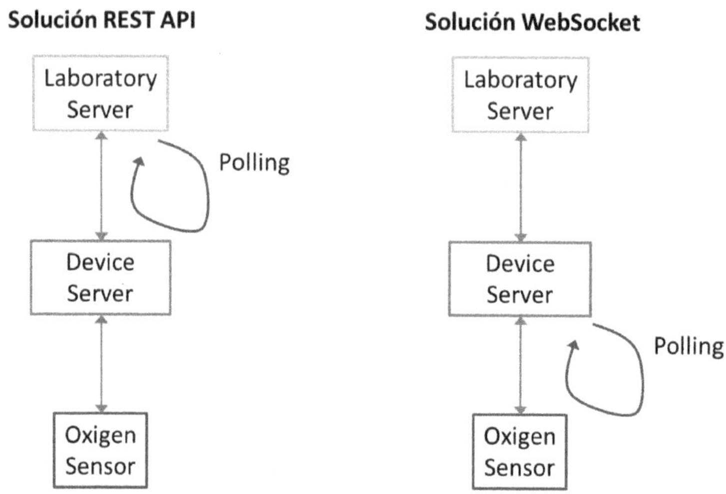

Figure 6.12. REST API and WebSocket solutions for communications in a remote experiment.

- **Standard implementation.** Implement the device server in a Raspberry-type microcontroller, and the lab server in a normal PC/server. In this case, the lab server can take on more complex tasks via deployment of WebLab-Deusto: tracking, federation, etc.

To summarise, the design team should take decisions in architecture and communications that will condition the quality of the remote laboratory. Basing oneself on the work of others, like WebLab-Deusto, is always a perfect starting point.

6.3 WebLab-Deusto as a Platform for Development of Remote Experiments

WebLab-Deusto and WebLabLib, along with other tools, constitute the only open platform that facilitates the task of implementing remote experiments and labs. This software initiative is coordinated by Pablo Orduña and Luis Rodríguez, founders of LabsLand.

WebLab-Deusto (described in Section 6.1) can be downloaded from GitHub (https://github.com/weblabdeusto), and in doing so the designer obtains all the characteristics described previously: decoupling, security, scalability, LMS connectivity, universality, tracking, queue management, federation, etc. There is the option of standard installation or a more complex personalised configuration.

Once WebLab-Deusto has been downloaded, the designer begins with the specific task of a remote experiment — in other words, begins to design the device server and the lab server.

The device server largely depends upon the experiment's hardware and the designer 's experience, bearing in mind that ultimately it will have to communicate with the lab server.

In the lab server is the interface, the logic of the experiment and communication with WebLab-Deusto and the device server. The WebLabLib library (https://github.com/weblabdeusto/weblablib) was created to assist designers in this task.

This library and other tools facilitate the designer's task so that instead of programming services they will only have to configure certain tools. In other words, the designer does not need command of the underlying technologies, but of the WebLabLib library. In any case, the designer should have IT knowledge, since this is not a trivial task.

For example, if the device server was designed in LabVIEW, then WebLabLib helps the designer to create the lab server and its communication. The same applies to EJS or the use of REST API and/or WebSockets.

The design of ultraconcurrent or pre-recorded remote experiments is a special case and is also supported by LabsLand.

LabsLand's objective via these tools is to promote the mass design of experiments that can be added to the federation of experiments in such a way that the educational community may access more of these, creating, within a collaborative economy, a federation of remote experiments that each centre contributes to and accesses.

6.4 Conclusion

The proposed architecture meets the main goals that were stated: decoupling, reliability, scalability and security. By decoupling the components and splitting the architecture into three layers, it is possible to develop and test each component separately. This offers various significant benefits. Experts (in hardware and software) can develop each layer independently, and the result is more likely to be reliable. Likewise, and though this has not been discussed in detail, such an architecture is also easier to scale to a higher number of simultaneous users, because it can be replicated with relative ease. Though there are different ways to do so, a simple method is to replicate the hardware and to deploy more instances of the processes. The WebLab-Deusto layer supports replicated laboratories, so once the remote laboratory is deployed properly the WebLab-deusto can balance its load automatically or even turn off instances of the laboratory that fail. This way, it is relatively straightforward to obtain high-quality laboratories, which can scale to an arbitrary number of users (with enough hardware replication) and can be highly reliable (because they fail less and because replication makes it possible to turn off failed instances).

Part

4

Innovation in Remote Experimentation

Chapter 7

Advanced Topics in Remote Experimentation

Chapter 7 addresses diverse, advanced aspects of remote experimentation. For example, a remote laboratory generates a considerable amount of data that, if analysed correctly, can improve the laboratory's performance by means of advanced tools. The purpose of the first section is to present the objectives known generically as Learning Analytics.

Furthermore, interfaces are the means by which students interact with the remote experiment, and their design will therefore affect the sensation of immersion and the utility and usability of the remote experiment. These two aspects will be addressed in the next two sections.

7.1 Adaptive Learning, Learning Analytics, Big Data, Online Tutorials and Remote Labs

Different studies highlight the importance of tutorials as part of the teaching-learning process, whether experimenting in the laboratory or studying in class. This importance does not depend on the local or remote nature of the experiment but on the teaching process itself.

In any case, in remote experimentation the online tutorial is fundamental, as the student will not be in the classroom surrounded by classmates and the teacher, but will be alone. How can queries be resolved?

Tutoring Systems are fundamental in online teaching, as are systems of automatic evaluation. In online teaching, this field is very active and has been developed via chatbots, videos, Intelligent Tutoring Systems (ITS), etc. How can this issue be addressed in the case of remote experimentation?

In VISIR for example, what happens if students are unable to assemble a circuit? Does VISIR help them? Do they send a message to a teacher or a classmate asking for help? Who sends the message? At the moment VISIR does none of this, considering that students themselves should analyse and resolve the problem. In this vein, Rodríguez-Gil *et al.* (2019) designed in his doctoral thesis a platform allowing the teacher to design a Tutoring System that in turn helps students when necessary. The student will ask the Tutoring System, and the latter will respond with more or less precision. The platform designed is interesting because it does not require technical knowledge on the part of the teacher who designs it (each teacher can design their own Tutoring System) and moreover the Tutoring System can be integrated within any environment, including a remote laboratory.

Figure 7.1 shows the TS designed for VISIR. The student asks the person on the right questions and the latter responds. The quality of the TS depends on the knowledge of the teacher who designs it.

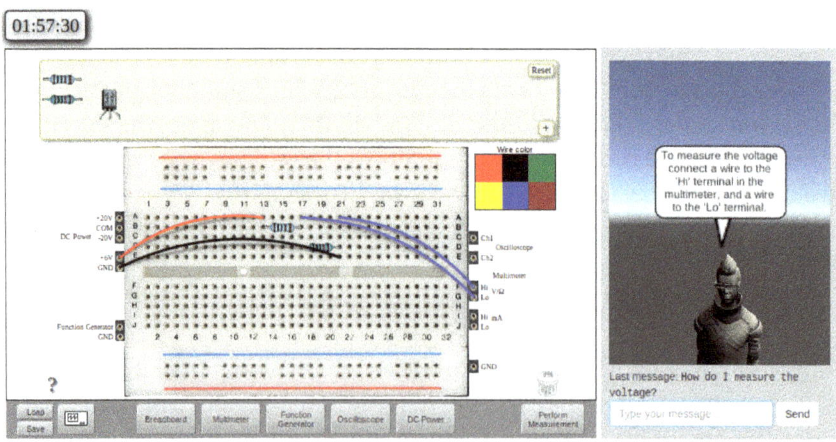

Figure 7.1. Intelligent Tutoring System with the VISIR remote experiment (Rodríguez-Gil *et al.*, 2019).

There are a multitude of TS, some described as intelligent — ITS — that are capable of embracing, to a greater or lesser extent, the learning scenario and in particular the students' evolution therein, giving them a kind of tutorial that will not only be requested by the student, but will even be proactive. It is not the goal of this book to discuss these types of tutorials, simply to indicate that they are crucial for remote laboratories and that as yet there has not been much development in this regard.

Another significant aspect in a remote experiment is automatic or online evaluation. Can a teacher know if the student has done the experiments indicated? Can the teacher know whether the result was correct? Currently, the teacher has to organise tests or evaluation methods separately from the remote experiment: exams, submission of documentation, tests, etc. But how do teachers know that the student has not copied the results that they have just been sent? How do they know the student has completed the programme executed in the robot? It is true that these questions are also pertinent in the real laboratory, but they are more delicate and urgent in the remote laboratory. Any automatisation of this process would be of great help to the teacher, and to the student too. Let us not forget that a quality evaluation should be summative and formative; in other words, it should provide a grade and explanations thereof.

Once again, in this field there has been little tangible development of use to the community. In the case of the VISIR remote laboratory, it is worth highlighting the works of (Romero, 2015) and (García Loro, 2018), and also that developed in (García-Zubía *et al.*, 2019). The latter work has a transversal approach and seeks to be a VISIR dashboard, while the other two works are directed towards evaluation alone.

The VISIR-DashBoard is based on the fact that the RLMS WebLab-Deusto stores all the actions performed by the students and the VISIR itself. In other words, everything is stored and can therefore be consulted and analysed.

Figure 7.2 shows how long a certain group of students has worked in VISIR. We can see there are two 'peaks', one of which may correspond to the more active students (further to the right), while the other corresponds to the normal or less active students (the latter is higher than the former).

It is also possible to analyse the information in detail for each student and each day, as shown in Fig. 7.3.

Finally, the teacher can also see how many different circuits the students have assembled (see Fig. 7.4).

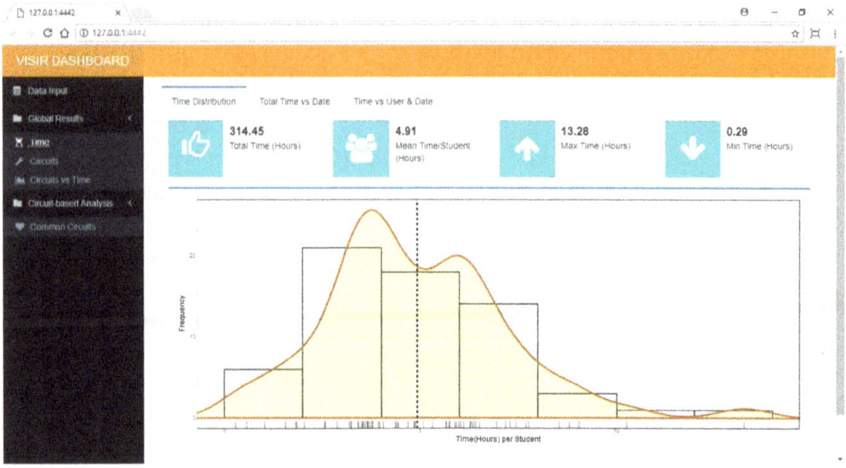

Figure 7.2. VISIR dashboard with student access.

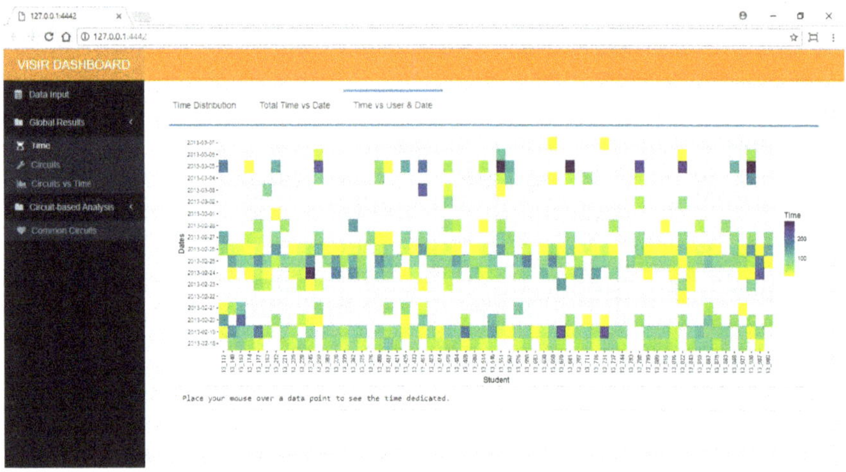

Figure 7.3. VISIR dashboard with hourly student activity.

Figure 7.4 can again be observed for each student and for each type of circuit and measurement according to Fig. 7.5: resistance, voltage and current.

While the Figs. 7.2–7.5 show more or less raw data, in Fig. 7.6 there is an element of analysis. Figure 7.6 shows each student according to two axes of coordinates. The horizontal axis shows the time devoted by each

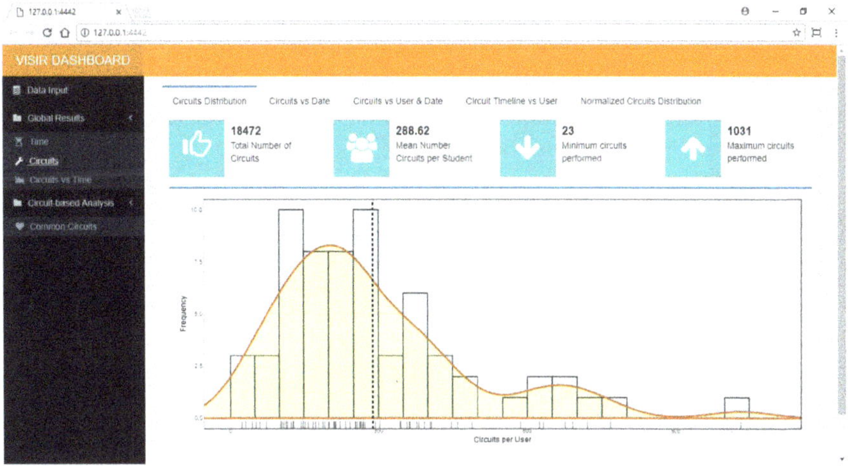

Figure 7.4. VISIR dashboard with student activity by circuit.

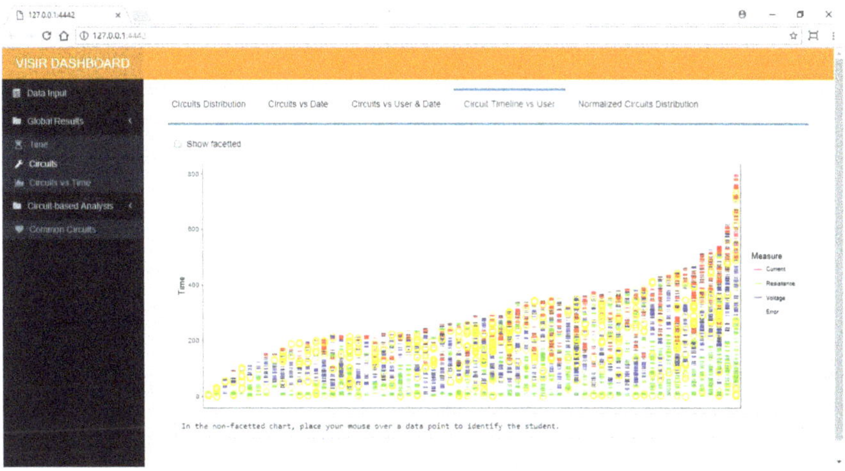

Figure 7.5. VISIR dashboard with the types of circuit created.

student up to 1000 minutes (approximately 16 hours), while the vertical axis indicates the number of different circuits assembled and measured, with a maximum of 200 circuits.

Thus, for example, if a student is in the top right-hand corner, it means they have done a lot of experiments and taken a long time; in other words, this is a good, hard-working student. If a student is in the bottom left-hand corner it does not mean they are bad, but that they have

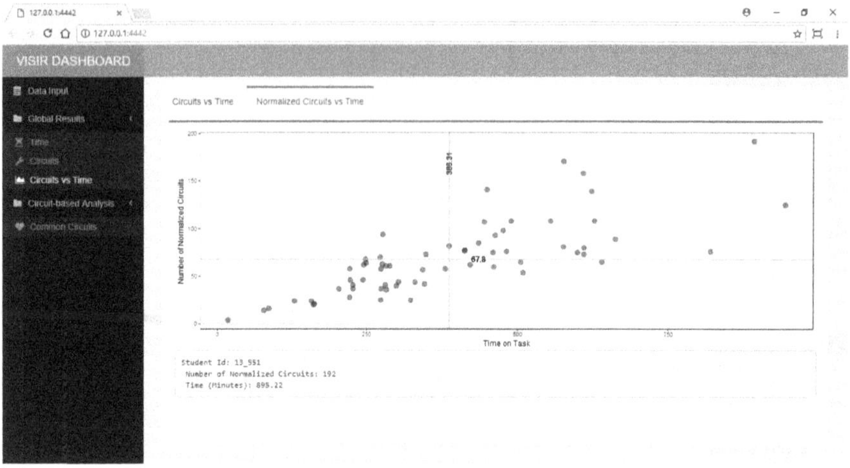

Figure 7.6. VISIR dashboard relating student, time and number of circuits created.

done little, in little time. A student may be regarded as not very good (or bad) if they are located in the bottom left quadrant, as in this case this means they have devoted a lot of time to masking few circuits, which might suggest that the student needs support, reinforcement or a tutorial (online or face-to-face). Brilliant students are to be found in the top left-hand quadrant, in which case the student has assembled and measured many circuits in a short time, which reveals interest (many circuits) and efficiency (little time).

It is not our aim to continue analysing this tool still under development, but it clearly offers many pedagogical elements, both of a teaching nature in the classroom and research-related.

There are important questions to be asked. Has the student learnt to assemble and measure an electronic circuit? Do they make fewer mistakes at the beginning than at the end? What kind of mistakes: conceptual or instrumental? Is it possible to identify when a student needs support in real time? Can formative evaluation be given to the student in real time?

Intensive work in this field has a lot to do with big data in general and with treatment and analysis of big data. The application of these and other more specific and educational techniques gives rise to what is known as Learning Analytics, LA. This is obviously a field with a major impact upon the teaching-learning process in the classroom and with particular relevance in online education, and specifically in remote experimentation.

As is customary in research, it is important to distinguish between research that provides applicable results and the rest, and in this case it is teachers themselves who should identify what is useful for them and/or their students.

Finally, adaptive leaning seeks to adapt and create technological tools to improve the student's learning process. For example, create Online Tutorials and design Adaptive Interfaces. If the experiment knows via Learning Analytics what students are doing and what errors they are committing when using the remote experiment, then one possible action is to adapt the interface to help students in the learning process (another was Online Tutoring). For instance, if it sees that a student is not powering an electric circuit correctly, then this can be provided in pre-assembled form. Or if the level of knowledge of buoyancy is seen to be low, then not all sensored signals will be shown on the interface, only those that appear useful for the student. This is currently a very active field of research.

7.2 Interfaces and Remote Laboratories

This section will address how the use of advanced technology in the design of interfaces can substantially modify the use and effectiveness of remote laboratories. The web interface is the user's port of access to the remote experiment, so any modification of the latter may affect its deployment in the classroom.

7.2.1 *Advanced interfaces: Augmented reality, virtual reality and haptic devices*

An interface with augmented reality allows the designer to superimpose images upon the webcam video to make visible to the user information that is not directly visible. For example, in Fig. 7.7, the water in a tank can be dyed with different colours to indicate its temperature. The reality of the video is 'augmented' with this colour.

Figure 7.8 shows the RRLAB interface (Cahue *et al.*, 2013) where the user has defined the model of a pendulum's behaviour. By clicking ON, the user can see the real pendulum and, superimposed, the simulated pendulum in a scenario of augmented reality. In this case, the user expects to see that the simulated and real pendulum are more or less superimposed. Given the use of real and simulated elements, this could well be regarded as a hybrid experiment.

Figure 7.7. Remote experiment interface with augmented reality (Vargas *et al.*, 2013).

Figure 7.8. Interface and augmented reality of RRLAB remote experiment.

Technologically speaking, a remote laboratory with augmented reality makes it necessary to combine the webcam image with that of a simulated environment to create a new learning context, as is shown in Fig. 7.9. A challenge for the designer is to avoid excessive 'augmentation' of the image of the experiment, as this over-information might confuse the user.

In another example, what is augmented is not something that it is difficult for the user to see —water temperature, electric current, etc. — but the actual experiment or scenario. Mejías (2012), for instance, is able to change the room in which a remotely programmed robot moves

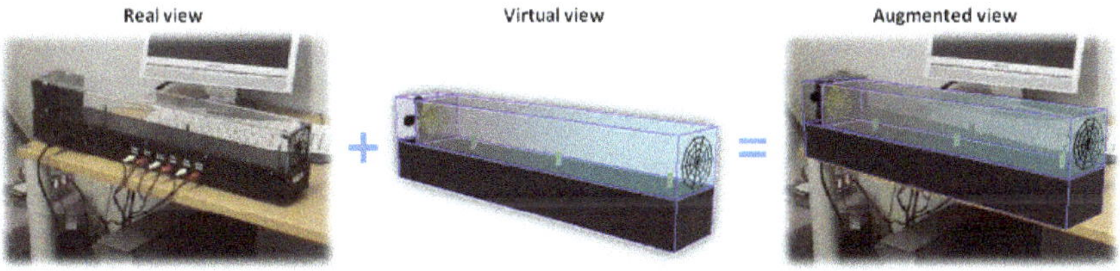

Figure 7.9. Structure of augmented reality.

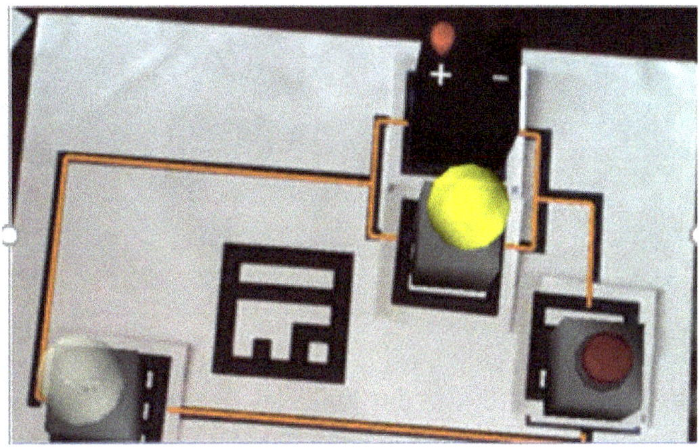

Figure 7.10. OE@FEUP experiment with augmented reality (Restivo *et al.*, 2014a).

(https://www.youtube.com/watch?v=4myvgl4Losc), which enriches the experiment at no further cost.

Restivo *et al.* (2014a, 2014b) suggest assembling an electric circuit using real QRCODE, which, when inspected by a webcam and software, produce a scenario of augmented reality where the student interacts with a simulated model (see Fig. 7.10).

According to Rodriguez-Gil *et al.* (2017), the previous two cases can be regarded as hybrid remote experiments (see Fig. 7.12), as they combine real with other simulated elements that interact within the remote experiment itself.

When we speak of virtual reality, we refer less to the interface itself than to its relationship with the user. The clearest example is a remote laboratory in *Second Life*, since in this case students do not move the mouse on their screens but move the hands of an avatar in the virtual world.

García-Zubía *et al.* (2010) developed a robotic remote experiment in *Second Life*. The user wrote a programme in C, then uploaded it to the

Figure 7.11. WebLab-Deusto robotic remote experiment in second life.
Source: https://weblab.deusto.es/website/second_lab.html.

robot and finally saw how the robot evolved: if it managed not to crash, if it followed the line, etc. The difference lay in the fact that the student interacted in the Second Life environment (https://weblab.deusto.es/website/second_lab.html) via his avatar, instead of using an interface with mouse and screen (see Fig. 7.11), which is not very active currently.

Game-Based Learning uses a game so that when interacting with it the student has to acquire and demonstrate certain competencies and skills. There are few remote experiences with this focus, but prominent among them, in terms of level and use, is Circuit Warz (available in STEAM) coordinated by Michael Callaghan of Ulster University (Callaghan *et al.*, 2012). In Circuit Warz, the student plays the role of an armed soldier who has to resolve problems where electronic circuits need to be designed (see Fig. 7.12).

The eLab3D (López *et al.*, 2015) laboratory developed by UPM (Spain), and shown in Fig. 7.13, takes the student into a virtual world where their avatar can assemble and measure electronic circuits. The design is technologically complex and requires knowledge in diverse fields, as it is necessary to combine reality (LabVIEW) and virtuality (Opensim) without dissonances.

On entering (see Fig. 7.14), students find themselves in an environment that reproduces a laboratory workstation where they can work

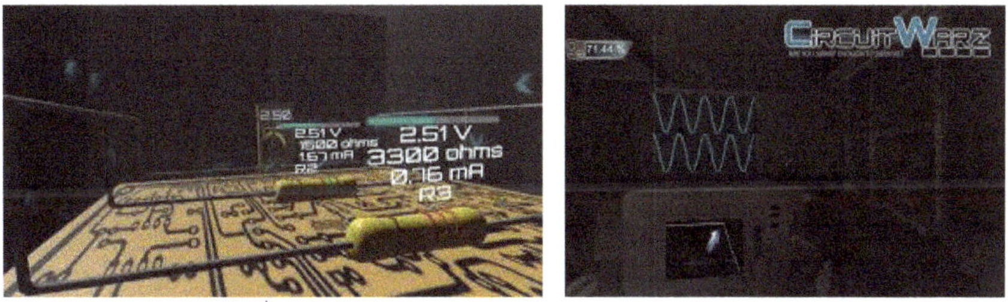

Figure 7.12. Circuit Warz game with remote experimentation.
Source: circuitwarz.com.

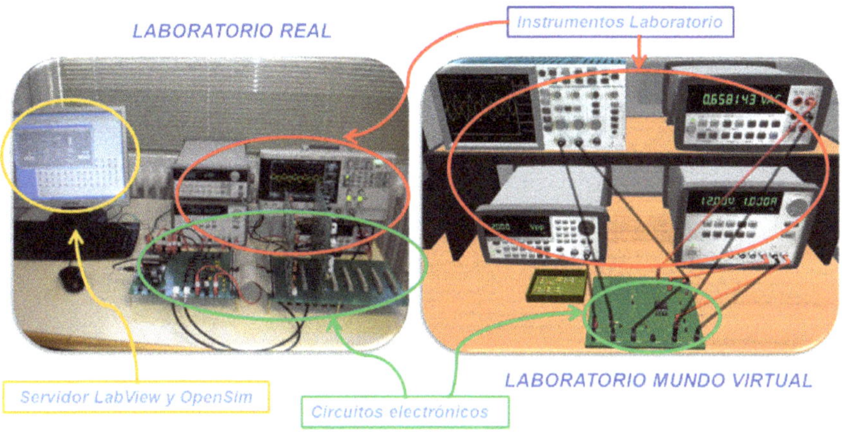

Figure 7.13. eLab3D remote experiment integrated in a virtual world.
Source: https://3dlabs.upm.es/.

(open drawers, move wires, etc.) as if it were real (https://www.youtube.com/watch?v=O0sxvTwvKzI). In this case, the student might wonder if what they are doing is real (real measures) or virtual (simulated), even if they are able to tell the difference.

However, in order to access a sophisticated and powerful system, in many cases it is necessary to install some kind of plugin or visualiser. Once again, this compromises the universality of the remote laboratory, which simply confirms that in remote experimentation power compromises universality and vice versa. Each designer and user must choose. In the case of 3DLabs, the message received by the user appears in Fig. 7.15.

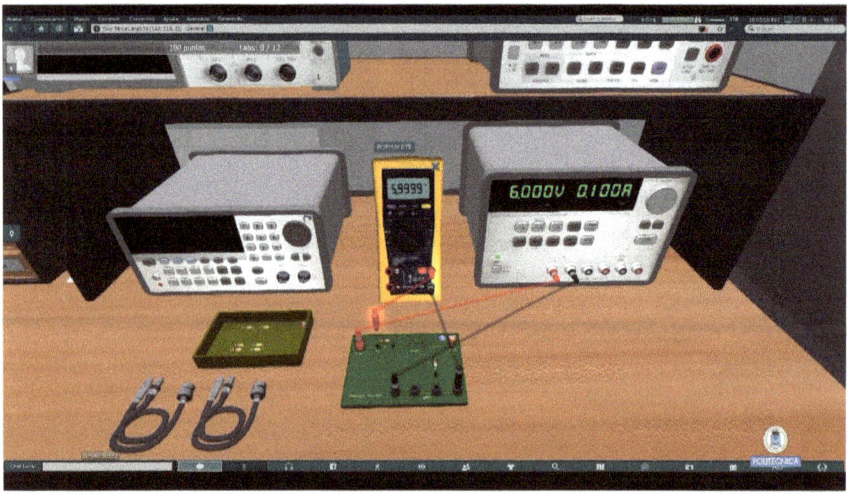

Figure 7.14. eLab3D student work surface.
Source: https://3dlabs.upm.es/.

Figure 7.15. eLab3D installation requirement.
Source: https://3dlabs.upm.es/.

The work carried out by the team led by Teresa Restivo in FEUP (Portugal) is extraordinary in the context of virtual technologies. Their remote and virtual laboratory offers various experiments related to mechanical engineering, materials technology, civil engineering, biomedicine, etc., and in these cases the use of augmented reality or other similar technologies helps students in their learning process (Rodrigues *et al.*, 2017). A special technique employed only by this team is the use of haptic devices. In this case, the user's experience is not enriched in the interface via colours that highlight the value of a force or tension, but by

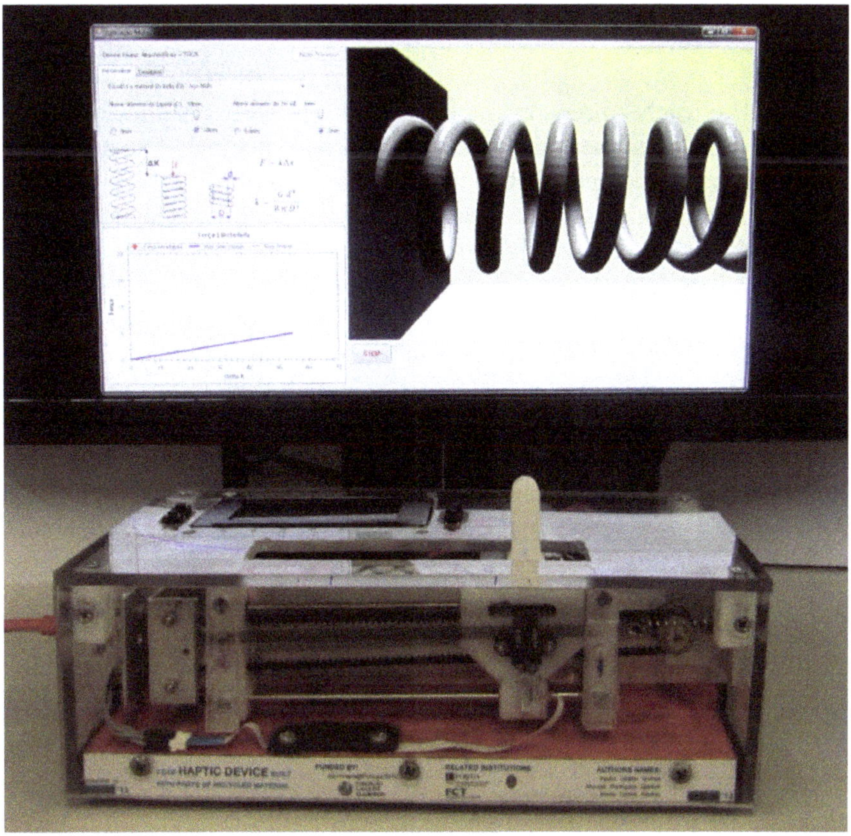

Figure 7.16. Haptic device of the OE@FEUP (Quintas *et al.*, 2013).

means of a real device — a haptic device — that is connected to the client's computer and which the student should have or manufacture with a 3D printer or with the downloadable FEUP models.

Figure 7.16 shows the interface and the haptic device of the OE@ FEUP laboratory according to Quintas *et al.* (2013).

For example, if the student wishes to increase the force applied to a steel bar to test its torsion, then the student does not use the mouse on the interface but pushes a plunger that resists the push, which makes the user's experience more realistic. Added to this is the integration of OCULUS glasses (www.oculus.com) and the use of Virtual Reality or VR.

The research efforts of this and other similar teams are leading the way into the future and showing others where opportunities lie.

7.2.2 *Interfaces and learning outcomes*

In an experiment in a hands-on laboratory, the interface is the actual student. The teacher prepares the experiment in two ways: they organise the laboratory equipment and give out a practice manual. The student is the fundamental element, but the teacher can condition the experiment. For instance, imagine that the teacher has set up a microscope with the sample positioned, centred, illuminated and focused on, and the student is asked simply to look through the microscope and write down whether or not they see a cell. Is the student really performing a scientific experiment? It is clear that the student has the privilege of seeing a cell, but does this constitute a learning experience? Have they investigated? Have they applied the scientific method?

As we have said, in a remote experiment, the student's hands and eyes are the interface, and therefore the design of this interface conditions the student's experience and learning. The designer, not the teacher, conditions or biases the development of the experiment, and consequently its effectiveness.

Not all experiments are similarly conditioned by the interface, since much depends on the nature of the experiment. For example, the basic interface of a microscope is very simple: it merely needs to allow the student to alter the magnification, move within and focus upon the sample. Depending on what they see, students make notes.

Another example is the LabsLand pendulum. The interaction is simple: raise and release the pendulum. The initial objective is to obtain the period of the pendulum, while the learning result will reside in knowing whether or not this depends on the angle of release of the pendulum. The question, in this section, is how to measure the period.

On the one hand (see Fig. 7.17), the interface provides a chronometer that can be used to see how long a cycle lasts or, better still, to see how long 10 cycles last and then divide that figure by 10, rendering the value more accurate.

The same clock can be used to measure 10 seconds and count the number of cycles during that time. The problem is that in this case the number of cycles is unlikely to be an exact figure, and therefore we may make a greater error.

Another option offered by the interface is to download the angular position at each moment onto an Excel file. Once the file is downloaded, the student can take a much more precise measurement.

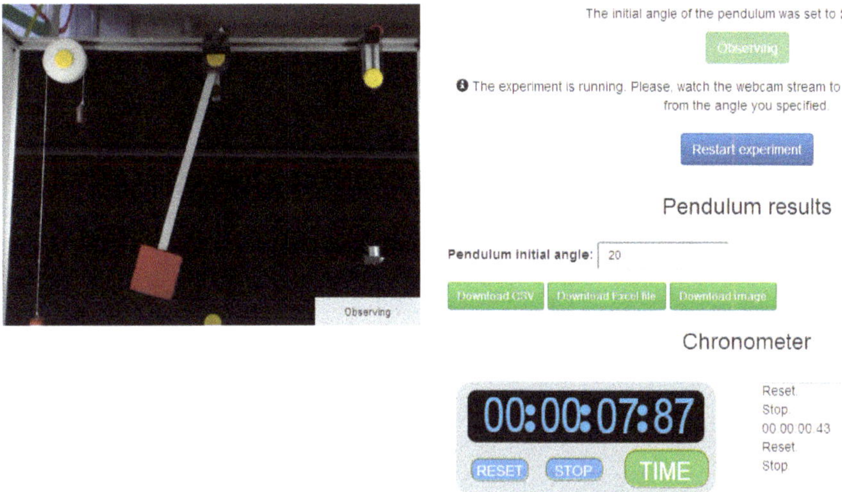

Figure 7.17. Interface and measurements with a LabsLand pendulum remote experiment. *Source*: www.labsland.com.

The table below shows the results obtained for each of the three methods for an angle of 20°.

	10 cycles	10 seconds	With excel	Calculated mathematically
Period	10/7.87 = 1.27 s	13/10 = 1.3 s	1.34 s	1.4 s

The three values are similar, and the closest to the calculated value of 1.4 s is the one that uses Excel. However, it is much more fun and easier for the student to use the chronometer and watch the pendulum. Students can even use their own watches.

In addition to the above, the remote laboratory, providing it is sensorised, could give the student the value of the period directly. In this case, the student would not be an active subject in the experiment but would simply raise the pendulum and press ON and the interface would show the pendulum period. Will there be a positive learning result? Does the degree of involvement in the experiment condition the student's learning? To what extent?

In this scenario, it is worth considering for a moment the experiment performed by Márquez-Barja *et al.* (2018). The experiment focused on the

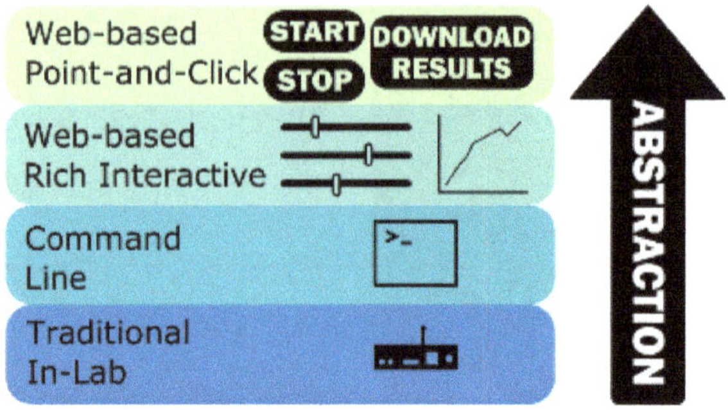

Figure 7.18. Abstraction level of a remote laboratory interface (Márquez-Barja *et al.*, 2018).

field of telecommunications and was offered to students via four interfaces, summarised in Fig. 7.18.

There are four different interfaces for the same experiment (see Fig. 7.19):

- **Web-based Point-and-Click:** The student chooses a series of parameters from a menu, and when they press start, the webpage transforms these into commands, the results of which are compiled later.
- **Web-based Rich Interactive:** By means of widgets, the student sets the inputs and sees on a graph and in real time the behaviour of the output.
- **Command line:** The hardware is controlled from a distance via screen commands.
- **Hands-on lab:** The student must be in front of the equipment to control it. Commands are generated manually, not via software.

The research question was which interface produces better learning results?

On the one hand, 'Naturally, the student responses indicate an increasing reduction in laboratory difficulty as abstraction increases, until a sharp corner is reached', but on the other hand, 'the interactivity offered by the user interfaces in support of self- assessment activities is lower'.

Figure 7.19. Different student interfaces (Márquez-Barja *et al.*, 2018).

The experiment focused on five questions, and analysis thereof was through student surveys (see Table 7.1).

Another option is prior analysis of the interfaces on two axes: passive–active and simple–complex. In this way, Fig. 7.20 shows that the most valued interface is that which, not being excessively complex, allows the student to control the activity. This may be the point of maximum immersion.

The conclusion is that students prefer the web-based Rich Interactive interface. This interface facilitates the work for them without any need to worry about the low level. This concern would be reduced even further by the Point-and-Click interface, but in this case students have the sensation of not controlling the experiment or that it is not real.

Table 7.1. Results of Márquez-Barja *et al.* (2018) questionnaire for different interfaces.

Totally agree+agree	Point-and-Click (%)	Rich Interactive (%)	Command (%)	Hands-on (%)
Was a helpful experience?	77	91	92	90
Impact of the user interface on the difficulty of the laboratory exercise.	22	91	63	47
Students' awareness that they were performing real experiments	68	90	76	89
Students' ability to self-assess progress during the course	45	91	75	89
Shows the eagerness of students to use the test facilities in the future	29	72	45	64

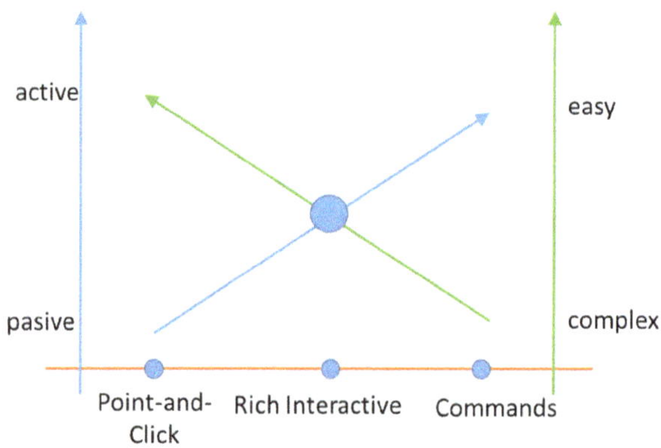

Figure 7.20. Analysis of interfaces.

It is also interesting that between controlling the equipment at a low level via commands or manually, students prefer the latter. They prefer the original hardware to the software copy.

The study by Marquez-Barja *et al.* (2018) is exemplary, because the scenario is simple and the statistical study is impeccable. This same experiment and other similar ones could be performed in other laboratories. For example, in the remote pendulum experiment, should we measure with the chronometer or without? With Excel? Receive the result directly? Is the graph useful?

And the same could be done with the buoyancy/Archimedes or the radioactivity experiment.

In the radioactivity experiment, a further question arises. In this case, the experiment is recorded and offered on a deferred time basis: the student configures the experiment (material, absorber, distance, time and repetitions), the RLMS searches for it in its video library and offers it in real time. This experiment takes a few minutes to perform, but could be offered to the student immediately — provided it is recorded. Is it worth doing this? Would it be worth giving all the tabulated results of radioactivity? In this case, it would no longer be pre-recorded, but Data Set.

Once again, it is the student or the teacher who has the answer, but the initial proposal generally comes from the designer, who may or may not have teaching experience.

To this debate over how much information the interface should provide and how abstract it should be is added the use of augmented reality. For example, if the water temperature rises in an experiment, is it enough to show this on a thermometer or is it better to colour the water virtually to highlight this? Is there a point at which the interface overwhelms the student with an excessive quantity or quality of information? This effect is multiplied with the use of virtual reality glasses and has already been addressed.

All these concepts may be observed as a sensation of immersion. So all the technologies and strategies described should aim to fulfil something that is not completely defined and is different for each student and experiment, but which everyone understands.

7.3 Usability and Usefulness in Remote Experimentation

As well as analysing the technical questions concerning remote laboratories and their positive effect on the teaching-learning process, one needs to know the student's opinion, what their sensations are. This analysis is carried out by means of more or less suitable satisfaction surveys, which in general are of poor quality since they are created by the development engineers and not by experts in the usability and usefulness.

There are various forms of expressing what usability and usefulness are, and, in general, the reader has a clear idea in this respect. Which is perhaps why Zhang's (2015) definition is interesting:

Usability or ease-of-use: 'the degree to which a person believes that using a particular system would be free from effort'.

Usefulness: 'the degree to which a person believes that using a particular system enhances his or her job performance'.

There are many surveys of this kind, but particularly deserving of mention, given the number of respondents, are those designed by UNED, NCSLab and WebLab-Deusto. These three questionnaires, and many others, usually include 10–20 questions, and frequently employed examples are the following: '[The system] has helped me to acquire the necessary competencies'; 'I would like to use [The system] again'; '[The system] was more useful than the hands-on lab'.

The NCSLab questionnaire (Hu *et al.*, 2016) basically refers to the remote laboratory of their own design which integrates 3D resources to control processes. The survey is divided into three blocks: technical issues, educational issues and added value. The Table 7.2 shows some of the 20 questions from the original questionnaire.

Table 7.2. NCSLab user satisfaction questionnaire (Hu *et al.*, 2016).

Technical issues	Educational issues	Added value
Is the experiment instructions helpful for you to carry out experiments quickly and correctly in NCSLab 3D?	Is the NCSLab 3D helpful to motivate your interest in studying?	Which one is more interesting for you to carry out experiments, in hands-on laboratories or in NCSLab 3D?
Is the NCSLab 3D convenient for you to carry out experiments?	Is the NCSLab 3D helpful for you to learn the knowledge from classrooms or books?	Which one (hands-on or NCSLab) can help you to learn more knowledge?
Is the NCSLab 3D website stable during experiments?	Is the NCSLab 3D helpful for you to master experimental skills?	Which one is more difficult for you to carry out experiments, in hands-on laboratories or in NCSLab 3D?
How satisfied are you with the results of your experiments in NCSLab 3D in general?	Do you hope to apply NCSLab 3D in other courses?	You will recommend your classmates to use the NCSLab 3D system for the experiments.

The results of the survey are positive in general. About 80% of students are generally satisfied with NCSLab, 84% say NCSLab has motivated them to study and 83% would like to use a remote experiment again, as 90% of them claim to have developed experimental skills with NCSLab. When students compare hands-on and NCSLab laboratories, in general, they do not express a preference for either, though they recommend their use to other students.

The UNED research team (Dormido *et al.*, 2011) has divided the survey into four parts: Learning value, Added value, Usability and Technology Function (Using Remote Labs in Education) as shown in Table 7.3.

In the global evaluation of the questionnaire, in no case is there less than 75% acceptance by students, which is a positive result.

Table 7.3. UNILabs user satisfaction questionnaire (Dormido *et al.*, 2011).

Learning value	Added value	Usability	Technology function
The lab enhanced my ability to understand the theoretical material in a new way.	I was able to develop a better understanding of how to control the systems I experimented with.	The Web-laboratory was easy to use and understand.	I did not miss any important information due to technology errors.
The lab helped me to visualize the concepts to be learned.	I was able to work through experiences in a way that would have not been possible in a traditional lab.	The documentation, tasks and tools presented within the web environment and related with the laboratories were easy to follow and find.	The software requirements did not pose a problem for me.
I gained as much information as I would from a lecture explanation.	The level of interactivity in the laboratory was adequate.	I did not miss any web feature or additional tool for performing my experimental activities.	The response time is good enough.
	The feedback from the teacher was useful.		I did not have any problem when accessing a remote laboratory.
	The collaboration with students was useful.		

The University of Deusto survey for WebLab-Deusto has evolved over time. Table 7.4 describes the 16-question survey in four areas (Usability, Usefulness, Immersion, and Satisfaction) and was a synthesis of various remote laboratory questionnaires. Table 7.5 includes the current proposal for a questionnaire, with nine questions in three areas. This second questionnaire benefitted from the collaboration of the Asistembe (http://asistembe2.iqs.edu/) educational team of IQS (Instituto Químico de Sarria, URL) in Spain, which is responsible for the statistical process that has yet to be published. In the shortened questionnaire, with an alpha Cronbach value of 0.8, one question per category has disappeared, as well as the entire Satisfaction category, since that information was redundant.

The results obtained during the first term of the academic year 2019–2020 are shown in Fig. 7.21. In all the questions, over 70% of the answers are five or higher in a seven-point likert scale. The Q1 (ease-of-use) and Q8 (usefulness) questions are important, as will be seen later, and

Table 7.4. WebLab-Deusto user satisfaction questionnaire.

Usability	Usefulness	Immersion	Satisfaction
The remote lab is easy to use.	The remote lab meets my requirements.	Using the remote lab feels like using a real lab.	The remote lab is enjoyable to use.
I can predict the result of using each element in the remote lab interface.	The remote lab helps me learn.	My interactions with the remote lab seem real.	I am satisfied with the remote lab.
The remote lab layout is sufficiently apparent so that help is not needed to use it.	The remote lab will help me pass the course.	When working on the remote lab I concentrate on the assigned tasks.	Using the remote lab stimulates learning.
The remote lab interface is easy to read.	The remote lab provides opportunities to practice knowledge.	I was engaged in using the remote lab.	The remote lab assists learners to have positive feelings about the course.

Table 7.5. WebLab-Deusto shortened user satisfaction questionnaire.

Usability	Usefulness	Immersion
Q1. The remote lab is easy to use.	Q8. The remote lab meets my requirements.	Q2. Using the remote lab feels like using a real lab.
Q6. I can predict the result of using each element in the remote lab interface.	Q4. The remote lab helps me learn.	Q3. My interactions with the remote lab seem real.
Q7. The remote lab layout is sufficiently apparent so that help is not needed to use it.	Q9. The remote lab will help me pass the course.	Q5. When working on the remote lab I concentrate on the assigned tasks.

in both cases exceed 80%. The highest valued question is Q4 (learning), with almost 95%, and the lowest valued is Q7 (interface), with 71%. In Q7, students confuse how difficult it can be to measure current or voltage drop with a multimeter with how difficult this task is rendered with the VISIR interface. Once again, the teacher should analyse the results carefully.

Meanwhile, researchers in other fields (software design, mainly) have identified usability as a field worthy of research in itself that has provided clear results. Of particular interest are the SUS, UMUX and UMUX-LITE questionnaires (https://measuringu.com/umux-lite/), which can be used in any environment and measure the usability of a computer application.

SUS is a 10-question questionnaire employing a seven-point Likert System to measure the usability of the system (System Usability Scale, SUS). This questionnaire was reduced to one of four questions (UMUX), two positive and two negative, which, for example, for the VISIR remote laboratory would be:

> *[This system's] VISIR capabilities meet my requirements.*
>
> Using *[this system]* VISIR is a frustrating experience.
>
> *[This system] VISIR is easy to use.*
>
> I have to spend too much time correcting things with VISIR *[this system].*

Finally, UMUX-LITE consists of just two positive questions, the first and the third from the list above. The first considers the usefulness of the tool,

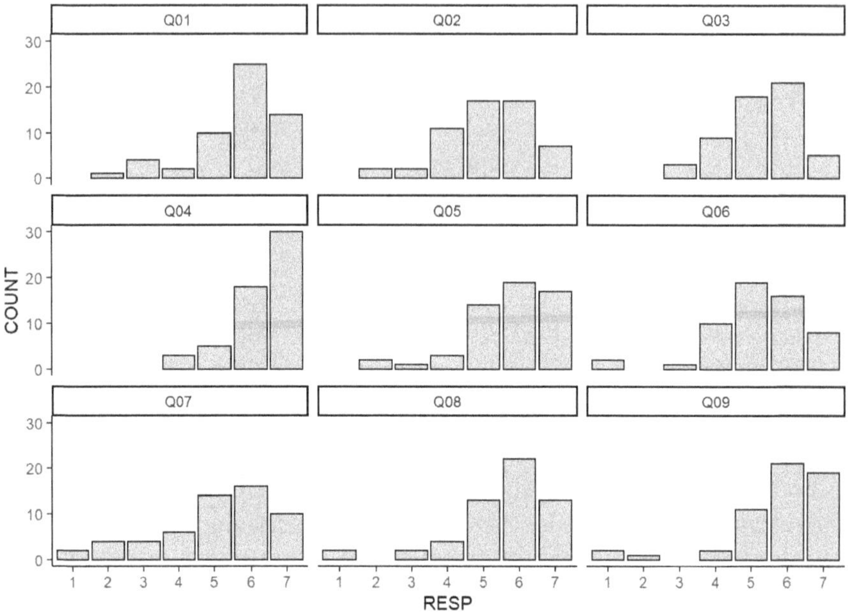

Figure 7.21. Results of statistical analysis of the WebLab-Deusto questionnaire in 2019–2020.

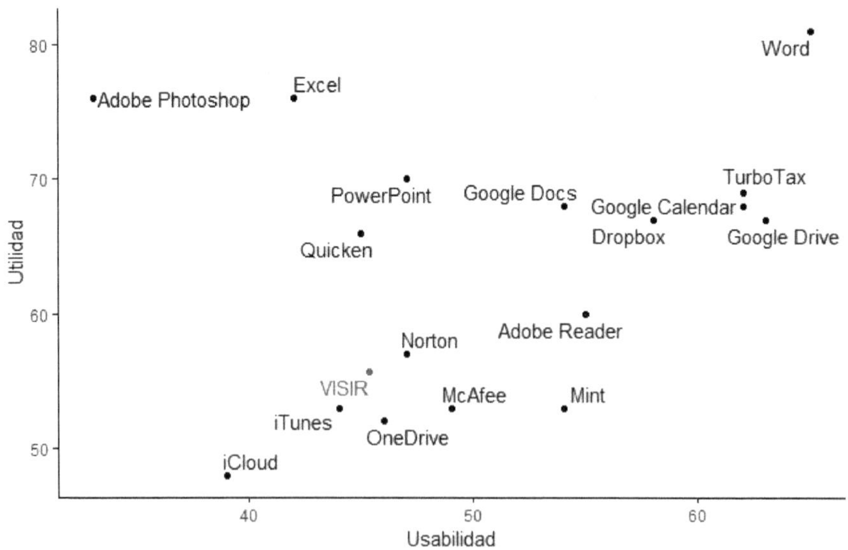

Figure 7.22. Analysis of usability vs usefulness for different environments.

and the second, the usability. Figure 7.22, obtained from J. Sauro in https://measuringu.com/umux-lite/, indicates that MS Word is the best IT product, while iCLOUD and iTUNES are the worst, according to the data available (https://measuringu.com/umux-lite/).

The University of Deusto, along with IQS, has analysed the VISIR remote experiment according to the J. Sauro's criteria. The value obtained exceeds 70, which should be regarded as good but not excellent. In any case, there is a difference for users between evaluating a tool that they have used, for example MS Word, and doing so with a tool on which they are going to be evaluated or which forms part of a subject. In the case that interests us, the student is unlikely to lean towards a positive evaluation as the tool forms part of the evaluation. Figure 7.22 shows how VISIR is situated in the lower left-hand area.

Data processing is a very rich field in this aspect and makes it possible to measure the UX (User eXperience). In the field of remote experimentation, it has not been used systematically (though the qualitative questionnaires have been used) and is therefore a field of work with considerable potential.

The next objective is for different research groups, and educational centres in particular, to use the proposed questionnaire, or others, and share the results, so the questionnaire can be evaluated on the basis of the highest number of surveys possible, to ascertain whether it can be further reduced without this resulting in loss of information. A long-term result would be the possibility of obtaining a single, unique indicator that rates the usability and usefulness of the remote experiment as a whole.

Chapter 8

Actors and Visions of Remote Experimentation

This chapter addresses remote experimentation from a different perspective, beyond education and technology. It aims to guide the restless reader in a different direction, in such a way that the results offer a different and more complete vision of remote experimentation.

8.1 Actors in Remote Experimentation

Remote experimentation involves different actors (see Fig. 8.1) with different needs and situations. From a conceptual point of view, these are:

- **Providers.** The team that provides the remote experimentation. This can be done via an IP, by means of an RLMS or a web page. There are different types of providers.
- **Consumers.** Users in general: students, teachers or anyone else.
- **Prosumers.** Consumers who are also providers.

Figure 8.1. Consumers and prosumers.

There are two kinds of providers. In the most common case, it is the designer of the experiment who offers the latter, and may do so alone or along with other experiments, in which case it would be a remote laboratory under an RLMS. There is also the case of LabsLand and similar situations.

LabsLand has developed its own remote experiments, so it is a basic provider, but also offers experiments from other providers, for example RexLab, (UFSC in Brazil) or Queensland (Australia). It does not matter to consumers whether the experiment accessed was designed by the team that publishes the webpage or by another. They are only interested in the conditions for access and the quality of the experiment.

It is currently normal for a consumer to access a remote experiment and use it in class, but they could also become a provider. Let us consider the case of the robot: the educational centre can organise with LabsLand — or another provider — the deployment in its installations of a copy of the remote robot in such a way that its students have priority access to this local robot. The educational centre can share its robot with other schools (or sell it), and vice versa, generating profits for all involved.

Teachers may also remotise an experiment in their centre, sharing it in LabsLand, so they can access those created by others. This may appear difficult at present, but the Czech company ISES (https://www.ises.info), for example, offers laboratory apparatus geared for the remotisation of physical experiments, and the USA company PASCO (www.pasco.com) sells instruments that publish their results online, and not only in a display, thereby facilitating the design of experiments that can easily be remotised. The hardware effort should be completed with the software effort necessary to 'take' that experiment to the platform provider (see Chapter 5). In this case, LabsLand offers modules that facilitate that connection. The technology is not yet complete, but it is incipient, and its development will be conditioned by the interest aroused by remote experimentation in industrial companies in the educational sector.

The above would represent a change of paradigm, since remote experiments would move from being designed by external teams in a top-down strategy to being designed bottom-up.

From the basic initial situation of provider-consumers, the addition of prosumers opens the scenario to other secondary actors who have already been relevant in other situations.

Educational authorities or networks are fundamental in order to deploy remote experimentation in schools and to analyse their educational value. Whenever remote experimentation does not require the real purchase of equipment, the Department of Education can purchase access to remote experiments and share them with educational centres. Public institutions and other institutional networks are responsible for detailed analysis of remote experimentation, with the aim of recommending the use of remote experimentation or modulating the latter in its application. For example, the Basque Government in Spain offers that possibility to public schools via the online Experimenta project.

The Australian Government made the considerable effort of taking the decision to introduce the first national network of remote experimentation following the report published within the LabShare project (Kostulski and Murray, 2010).

Moreover, educational centres, public and/or private, could assume the design, deployment and sharing of remote experiments, as it would seem very difficult for one centre to decide on its own to set up a remote experiment and search online for others with whom to share it.

It is time for educational institutions to assume a leading role that should shape the immediate future of remote experimentation. To the contribution of the administrative educational institutions should be added the participation of faculties of education, as these are training the teachers of the near future.

The primary mission of science museums is the active diffusion of science, either by receiving students or taking scientific activities and initiatives into the classroom. By going to a museum, students can observe experiments that are rarely available in their schools; if these experiments were remote, then students could access them anytime, anywhere. Science museums could be excellent allies of educational institutions in the effective deployment of remote experimentation, as they can function as providers or containers. It should be noted that science museums usually employ professionals in education and technology for the use and maintenance of the museum.

Publishing houses also play a significant role in remote experimentation. They have now digitalised most of their material, which is available on tablets. A large number of schools are adopting the tablet as a vehicle of content and activities in the classroom. These materials do not include remote experiments, and at most offer virtual laboratories or simulators. Whenever remote experimentation is digitalised, publishing houses could integrate it within their content, significantly increasing its value in the classroom. Remote experimentation could be an added value in a highly competitive field, and its development would involve a major technological effort, as we would not be speaking of one group of students or one school, but thousands of them, and the change of scale is therefore evident.

Suppliers of educational material regard remote experimentation as both a risk and an opportunity. If a company that sells an inclined plane offers the educational centre the possibility of remotising it, its value for the centre clearly increases, as the latter can share the resource with others centers. At the same time, however, this reduces the company's business; or not, as this may be an opportunity. The ISES Company is now offering this service, and some, like EDIBON (www.edibon.com), provide it on demand and at a very high cost. PASCO in the USA and ALECOP (www.alecop.com) in Spain are two companies that clearly exemplify this situation.

There are more and more distance training and learning companies and educational centres, as, for years, they have been using the Internet as a trampoline to reach an increasing number of students, be it in formal or non-formal education, MOOCs or classic courses. These companies do not often address training in technologies that are in great demand by companies such as microcontrollers, programmable automata (PLC) or any other programmable equipment like industrial robots, as students cannot complete effective practical training with simulators and virtual laboratories. Training companies could significantly broaden their options if they incorporated other remote equipment as a lever to increase and improve their training repertoire.

Noteworthy within this scenario is Brinson's (2015) work, as it poses the following problem (already mentioned) for the USA 'If an institution is offering postsecondary education through distance or correspondence education to students in a State in which it is not physically located, the institution must meet any State requirements for it to be legally offering postsecondary distance or correspondence education in that State'. The solution, he says, lies in the fact that 'The need for research related to the effectiveness of non-traditional learning environments, including the NTL, is being recognized by both educators and policy makers'. Brinson indicates that the answer is crucial for the development of more and better distance training qualifications in engineering, for example, in university studies. This question can be analysed by each country and university.

In short, for the development of remote experimentation, the community of researchers needs to be joined by other actors such as policy makers, educational institutions, science museums or distance training companies. Only collaboration between all these actors will produce a sustainable ecosystem of remote experimentation.

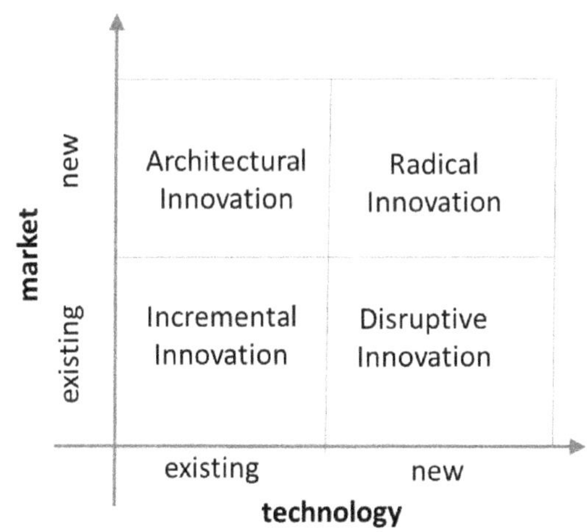

Figure 8.2. Classification of innovation.

Source: https://techblog.constantcontact.com/software-development/types-of-innovation/.

8.2 Remote Labs and Innovation

This section addresses remote experimentation from different angles than is usually the case, but always with innovation as protagonist, whether it be social, commercial, educational, etc.

Although there is clear and reputable evidence to support the claim that remote experimentation is an example of educational innovation, still to be analysed is whether this is an instance of incremental or disruptive innovation, as defined by Bower and Christensen (1995). This discussion recalls all the elements already addressed.

This book does not seek to clarify the differences between incremental and disruptive or radical innovation, but rather to use both concepts to situate remote experimentation in its context and to highlight its social, educational and technological potential.

Innovation is incremental when something is added to existing design to improve its market presence. Innovation is disruptive when, based on an existing product, a new one is created that makes it possible to penetrate a different market. This new product generally appears when new technology is applied, which will be disruptive. Let us see how remote experimentation reacts to this dichotomy (see Fig. 8.2).

On the one hand, it seems clear that the 'application of the Internet' to scientific experimentation has become a disruptive product. The technological effort is complex and considerable and is a clear example of Technology Enhanced Learning (TEL) but is it consolidated in the educational community? Has it increased the market? Is it a disruptive or a radical innovation?

A pertinent question is: how does the teacher address innovation, be it incremental or disruptive, in the classroom? It is not that teachers do not understand or seek the advantages of remote experimentation, but that they do not yet fully trust it. In order to understand this, it should be noted that there is the typical profile of the innovative and experimental teacher, but also the conservative profile, who does not want to risk using new tools that might compromise their way of teaching and their results. So, it is more important to understand and resolve teachers' doubts and listen carefully to their questions than to explain over and over again the advantages of remote experimentation. The designer is an important actor, but is far from the only one.

The technology of remote experimentation is mature and capable of offering, in a trustworthy manner, a sufficient number of remote experiments to prove their academic value, not to researchers but to real teachers in their classrooms. To this end, remote experiments need to be supported by a company or institution that is strong enough to immediately generate that trust. To this should be added the question of sustainability, which will also be supported by the agents that generate that trust, and vice versa.

One needs to emphasise, however, that remote experimentation is not being applied on a vast scale in educational centres, and is not proving successful. Why? Will it triumph in the future?

Remote experimentation will reach classrooms courtesy of teachers, and will remain there via the involvement of management. Backing and promotion by ministries and foundations will also be crucial.

Implicit in the above is the need to take educational, technological, economic and social risks, and to do so effectively in a specific period of time, and not as a non-stop research and innovative activity. A school is not an innovation centre, but it may be a centre for the application of innovation. The risks assumed and overcome should become real educational products.

In this respect, and continuing with the theme of this section, it makes sense to recall Umberto Eco, his semiotic theory and 'redundancy bands' (Eco, 1968). Umberto Eco points out that in order to introduce a new concept, remote experimentation, for example, we should include a channel that is well known to the end-user. If something is disruptive, it should be wrapped in familiar 'paper'.

If one accepts the disruptive nature of remote experimentation, what is its redundancy band and for whom should it be redundant? In the first place, in this approach the Internet is the protagonist, since 20 years ago it was a disruptive element, but now it can be used as a redundancy band of remote experimentation. The Internet, or digitalisation in general, can be the 'friend' that helps to popularise remote experimentation. The same role can be played by smartphones, where the Internet and Wi-Fi are consolidated technologies. Thus, what was a cause of surprise at a particular moment in time is now a source of security for users.

Students, perhaps more so still than teachers, are already accustomed to the digital environment, where many actions are performed digitally, via the Internet, apps and computer interfaces. For them, digital access to a real fact is not a source of concern or an epistemological doubt, but just another everyday action. In this sense, remote experimentation appears to be trapped in the digital divide, or is an obvious example of the latter.

In line with the previous reflection, it is worthwhile noting that Jane Hart (www.c4lpt.co.uk) compiled the 100 most popular tools in education, the first ones being YouTube, Google search, Twitter, etc. The first exclusively educational tool to appear is Moodle, in 35th position. Perhaps this is because teachers feel comfortable innovating in the classroom if they use familiar tools that constitute their redundancy band.

When speaking of other media in RE, it is worth including another lateral element in this section. In the 1960s, McLuhan introduced Laws of Media (McLuham, 1988), and his slogan was 'the medium is the message'; in other words, that changing the medium meant changing the message. Remote experimentation is a clear case of change of medium in the 'global village': if we move laboratory experiments to the Internet, then the experiments themselves change with McLuhan's approach. But how do they change? Can this change be analysed?

This analysis can be performed using McLuhan's tetrad, which analyses the new medium (or product, or technology, etc.) from four perspectives: Enhance–Reverse–Retrieve–Obsolescence.

Enhance	The technology must enhance some capability of the person. The medium is an extension of the person.
Retrieve	The result is a retrieval of some earlier service or capability that was lost and is now brought back into play in a new form.
Obsolescence	What is pushed aside and made obsolescent.
Reverse	If the new media is taken to extreme what will result that reverses the original characteristics of the media when it was first introduced.

The idea shows the tetrad and how it has been used to categorise the Internet, Twitter, Facebook… (https://mindbeforeyoumine.com/2010/07/12/mcluhans-tetrad-applied-to-internet/). Can the same be done for remote experimentation?

Continuing with this digression on the subject of remote experimentation, what award-winning photographer Joan Fontcuberta says makes a lot of sense. He comments that nowadays the function of a good photographer is not so much to take good photos as to choose the best: to be a good collector of images is to be a good photographer. In this vein, Joan Fontcuberta (www.fontcuberta.com) presented an exhibition of photos of sunsets taken by anonymous individuals and posted in Flickr. Fontcuberta studies how the democratisation of means of cultural production affects photography.

More specifically, the GoLab Project, as it was financed by the EU, was urged not to design virtual remote laboratories, but to look for existing ones, and order and evaluate them academically, in other words, codify, use and review them. Ultimately, are possible remote experiments as infinite as sunsets? Are basic and well-known experiments in each scientific area already remotised? Can they be discovered and displayed on the GoLab web?

Fontcuberta also addresses a fact closely related to remote experimentation and suggests that 'all photography is a fiction that is presented as true' ... 'what is important is how the photographer uses it, what purposes it serves'. Earlier, we presented two types of rather special remote experiments: those that were a set of data and those that were a set of recordings (both are very similar). Does their recorded/photographic nature distance them from their educational purpose? Does it make sense to take more, or is it also logical to enjoy those already taken? What element makes an already recorded experiment relevant for a student? Is it necessary to repeat it again every time for each student?

This section has obviously used theories from other areas to analyse remote experimentation from their perspective in order to enhance and situate the field of work. Remote experimentation is not restricted to technology and education, but is complex and rich.

Summary: The Ten Commandments of Remote Experimentation

By way of summary and in slightly more light-hearted fashion, here is a series of commandments for teachers:

- **Do not allow designers to tell you what is good for you:** Learn to tell them what you expect from the remote experiment to be used in the classroom.
- **Love IT services more than your remote experiments:** The remote laboratory in question should never compromise the educational centre's IT services. It is very difficult, if not impossible, for technical staff to follow educational criteria. Before doing anything, speak to them, and get on well with them!
- **Be professional when choosing the remote laboratory:** You must be sure that the remote experiment is going to work, you must accept only professional remote laboratories that are sustainable and of the highest quality. You do not need

explanations when something does not work in the classroom, you need it to work always (or almost always).

- **Think about the curriculum and you will succeed:** The remote experiment in the classroom should address a certain objective or competence of your course or subject, whatever the type. For example: understand the law of gravity, or practise English, or be innovative.

- **The curriculum is the teacher's, not the designer's:** Do not allow the person who designed and assembled a remote experiment to give you academic lessons if you do not agree with them. The experiment must convince you, you do not have to convince anybody of your knowledge.

- **Control the tool and not vice versa:** A remote experiment can change the way of teaching, and this will require time and reflection on your part, but you should never be far behind the tool.

- **Protect your students from your sense of innovation:** Do not add it without reason. Do not keep doing the same old thing in the classroom before adding your final innovative touch; that will tire and distract the students. Choose what is good, in your view, for the student and the subject, and apply it.

- **Students are not lab rats:** Do not use remote experiments to collect data from students in indiscriminate fashion, analyse and then publish it in journals. Remember the Ethics Committee.

- **The experiment should help, it should not in itself be a challenge:** In some cases, the interfaces of remote experiments are so complicated that they cancel out their educational effect. The student devotes more time to use than to utility, which devalues both the experiment and the teacher that introduced it.

- **Students are digital natives, all of them!** For a student, digital information is as real as analogical or natural information. Internet is real for them, try to create a quality and immersive experience, and do not worry about whether or not students are immersed, they are.

References

Agarwal, D. and R. Bair (2004). 'National collaboratories horizons, report of August 10–12 2004, national collaboratories programa meeting', in *the U.S. Department of Energy office of Science.*

Alves, G., A. Marques, C. Paiva, P. Nogueira, P. Guimarães, R. Couto, L. Cherem, V. Borba, G. Ferreira, S. Koch, and A. Pester (2013). 'A remote lab for projectile launch experiments: Professional and academic perspectives', in *Proc. 10th REV Conference*, pp. 1–8.

Alves, G. R., A. Fidalgo, A. Marques, C. Viegas, M. C. Felgueiras, R. Costa, N. Lima, M. Castro, G. Díaz-Orueta, E. San Cristobal Ruiz, F. García-Loro, J. García-Zubía, U. Hernández-Jayo, W. Kulesza, I. Gustavsson, A. Pester, D. Zutin, L. Schlichting, G. Ferreira, D. de Bona, J. B. da Silva, J. B. Alves, S. Biléssimo, A. Pavani, D. Lima, G. Temporão, S. Marchisio, S. Concari, F. Lerro, R. Fernández, H. Paz, F. Soria, N. Almeida, V. de Oliveira, M. I. Pozzo, and E. Dobboletta (2016). 'Spreading remote lab usage a system — A community — A Federation', in *Proc. 2nd International Conference of the Portuguese Society for Engineering Education (CISPEE)*, pp. 1–7.

Angulo, I., L. Rodríguez-Gil, and J. García-Zubía (2018). 'Scaling up the lab: An adaptable and scalable architecture for embedded systems remote labs', in *IEEE Access*, Vol. 6, pp. 16887–16900.

Auer, M., A. K. M. Azad, A. Edwards, and T. De Jong, Eds. (2018). *Cyber-Physical Laboratories in Engineering and Science Education*, Springer Ed., p. 434.

Azad, A. K. M., M. E. Auer, and V. J. Harward Eds. (2012). *Internet Accessible Remote Laboratories: Scalable E-Learning Tools for Engineering and Science Disciplines*, IGI Global, p. 645.

Bagnasco, A. (2007). Presentation in meeting '*International workshop in remote experimentation*', University of Deusto, Bilbao (Spain).

Bohus, C., L. A. Crowl, B. Aktan, and M. H. Shor (1996). 'Running control engineering experiments over the internet', in *Proc. IFAC* 1996, pp. 2919–2927.

Bower, J. and C. Christensen (1995). 'Disruptive technologies: Catching the wave', in *Harvard Business Review*, January–February pp. 43–53.

Brinson, J. (2015). 'Learning outcome achievement in non-traditional (virtual and remote) versus traditional (hands-on) laboratories: A review of the empirical research', in *Computers & Education*, Vol. 87, pp. 218–237.

Cahue, E., A. Ferrer, G. Ruiz, D. Iñiguez, and A. Tarancon (2013). 'RRLAB: Remote Reality Laboratory to teach mechanics in schools', in *IT Innovative Practices in Secondary Schools: Remote Experiments*, University of Deusto Ed., pp. 119–140.

Callaghan, M., K. McCrory, J. Lopez-Losada, J. Harkin, and S. Wilson (2012). 'Circuit Warz, the games; collaborative and competitive game-based learning in virtual worlds', in *Proceedings9th REV Conference*, pp. 75–79.

Corter, J. E., S. Esche, J. V. Nickerson, and C. Chassapis (2007). 'Constructing reality: A study of remote, hands-on, and simulated laboratories', in *ACM Transactions on Computer-Human Interaction*, Vol. 14, Nr. 2, Article 7.

Dalton, B. and K. Taylor (1998). 'A framework for Internet robotics', in *Proc. of IROS '98 Workshop 'Robots on the Web'*, pp. 15–22.

Dormido, S., J. Sánchez-Moreno, H. Vargas, L. de la Torre, and R. Heradio (2011). 'UNED Labs: A network of virtual and remote laboratories', in *Using remote labs in education*, University of Deusto Ed., pp. 253–270.

Dormido, S. (2004). 'Control learning: present and future', in *Annual Reviews in Control*, Vol. 28, Nr. 1, pp. 115–136.

Dziabenko, O. and J. García-Zubía, Eds. (2013). *IT Innovative Practices in Secondary Schools: Remote Experiments*, Ed. University of Deusto, p. 352.

Eco, U. (1968). *La estructura ausente. Introducción a la semiótica.* Ed. Lumen.

Esquembre, F. (2004). 'Easy Java simulations: A software tool to create scientific simulations in Java', in *Computer Physics Communications*, Vol. 156, Nr. 2, pp. 199–204.

Feisel, L. and G. Peterson (2002). 'A colloquy on learning objectives for engineering education laboratories', in *Proc. of the American Society for Engineering Education, ASEE* 2008, p. 12.

Feisel, L. and A. Rosa (2005). 'The role of the laboratory in undergraduate engineering education', in *Journal of Engineering Education*, Vol. 94, Nr. 1, pp. 121–130.

Fernandes, H., R. Neto, T. Pereira, I. Borges, G. Amarante-Segundo, S. Balula, R. Figueiredo, and J. Pereira (2013). 'World wide remote distributed experiment on E-lab' in *Proceedings Experiment@ International Conference, exp.at* 2013, pp. 212–214.

Ferreira, J. M. and A. M. Cardoso (2005). 'A moodle extensión to book online labs', in *International Journal of Online Engineering, iJOE*, Vol. 1, Nr. 2, pp. 1–4.

Fidalgo, A. V., M. Gericota, D. Barataud, G. Andrieu, R. De Craemer, M. Cristea, A. Benachenhou, M. Ankrim, K. Bouchlaghem, and P. Ferreira (2014). 'The EOLES project remote labs across the Mediterranean', in *Proc. of the 11th REV Conference 2014*, pp. 211–216.

Fjeldly, T. A. and M. S. Shur (2003). *Lab on the Web*, John Wiley & Sons, pp. 239.

Froyd, J. E., P. C. Wankat, and K. A. Smith (2012). 'Five major shifts in 100 years of engineering education', in *Proceedings of the IEEE*, Vol. 100, Special Centennial Issue, pp. 1344–1360.

Galán, D., D. Chaos, L. de la Torre, E. Aranda-Escolástico, and R. Heradio (2019). 'Customized online laboratory experiments. A general tool and its application to the furuta inverted pendulum', in *IEEE Control Systems Magazine*, Vol. 39, Nr. 5, pp. 75–87.

García Loro, F. (2018). '*Evaluación y aprendizaje en laboratorios remotos: Propuesta de un sistema automático de evaluación formativa aplicado al laboratorio remoto VISIR*', PhD. Thesis.

García-Zubía, J. and G. Alves, Eds. (2011). *Using Remote Labs in Education. Two Little Ducks in Remote Experimentation*, Ed. University of Deusto, p. 465.

García-Zubía, J. and L. Gomes, Eds. (2006). *Advances on remote Laboratories and E-learning Experiences*, Ed. University of Deusto, p. 310.

García-Zubía, J., D. López-de-Ipiña, P. Orduña, U. Hernández, and I. Trueba (2006). 'Questions and answers for designing useful WebLabs', in *International Journal of Online Engineering*, Vol. 2, Nr. 3, p. 6.

García-Zubía, J., I. Angulo, U. Hernández, and P. Orduña (2008). 'Plug&Play remote lab for microcontrollers: WebLab-DEUSTO-PIC', in *Proc. of European Workshop on Microelectronics Education, EWME 2008*, p. 2.

García-Zubía, J., P. Orduña, D. Lopez-de-Ipiña, and G. Alves (2009a). 'Addressings impact in the design of remote laboratories', in *IEEE Transactions Industrial Electronics*, Vol. 56, Nr. 12, pp. 4757–4767.

García-Zubía, J., U. Hernandez, I. Angulo, P. Orduña, and J. Irurzun (2009b). 'Acceptance, usability and usefulness of WebLab-Deusto from the students point of view', in *International Journal of Online Engineering*, Vol. 5, Nr. 1, pp. 1–7.

García-Zubía, J., J. Irurzun, P. Orduna, I. Angulo, U. Hernandez, J. Ruiz, M. Castro, and E. San-Cristobal (2010). 'SecondLab: A remote laboratory under second life', in *International Journal of Online Engineering*, Vol. 8, Nr. 4, pp. 34–39.

García-Zubía, J., J. Cuadros, S. Romero, U. Hernandez-Jayo, P. Orduña, M. Güenaga, L. González-Sabate, and I. Gustavsson (2016). 'Empirical analysis of the use of the VISIR remote lab in teaching analog electronics' in *IEEE Trans. on Education*, Vol. 60, Nr. 2, pp. 149–156.

García-Zubía, J., J. Cuadros, V. Serrano, U. Hernández, I. Angulo-Martínez, A. Villar, and G. Alves (2019). 'Dashboard for the VISIR remote lab', in *Proc. 5th Experiment@ International Conference, exp.at 2019*, pp. 42–46.

García-Zubía, J. (2003). 'Laboratorios virtuales vía web', *Keynote at Congreso Latinoamericano de Enseñanza de la Informática, CLEI 2003.*

Gomes, L. and S. Bogosyan (2009). 'Current trends in remote laboratories', in *IEEE Trans. Industrial Electronics*, Vol. 56, Nr. 12, pp, 4744–4756.

Gröber, S., M. Vetter, B. Eckert, and H. J. Jodl (2007). 'Experimenting from a distance — remotely controlled laboratory (RCL)', in *European Journal of Physics* Vol. 28, Nr. 3, pp. 127–141.

Gustavsson, I., K. Nilsson, J. Zackrisson, J. Garcia-Zubía, U. Hernandez-Jayo, A. Nafalski, Z. Nedic, O. Gol, J. Machotka, M. I. Pettersson, T. Lago, and L. Hakansson (2009). 'On objectives of instructional laboratories, individual assessment, and use of collaborative remote laboratories', in *IEEE Transactions on Learning Technologies*, Vol. 2, Nr. 4, pp. 263–274.

Harward, V. J., J. Del Alamo, V. Choudhary, K. DeLong, J. L. Hardison, S. R. Lerman, J. Northbridge, D. Talavera, C. Varadharajan, W, KYehia, and D. Zych (2004). 'iLab: A scalable architecture for sharing online experiments', in *Proc. International Conference on Engineering Education*, Vol. 9, pp. 9–43.

Harward, V. J., J. del Alamo, S. R. Lerman, P. H. Bailey, J. Carpenter, K. DeLong, C. Felknor, J. Hardison, B. Harrison, I. Jabbour, P. D. Long, T. Mao, L. Naamani, J. Northridge, M. Schulz, D. Talavera, C. Varadharajan, S. Wang, K. Yehia, R. Zbib, and D. Zych (2008). 'The iLab shared architecture: A web services infrastructure to build communities of Internet accessible laboratories,' in *Proceedings IEEE*, Vol. 96, Nr. 6, pp. 931–950.

Henke, K., T. Vietzke, H. D. Wuttke, and S. Ostendorff (2016). 'GOLDi — Grid of Online Lab Devices Ilmenau', in *International Journal of Online Engineering, iJOE*, Vol. 12, Nr. 4, pp. 11–13.

Henke, K., R. Hutschenreuter, H. D. Wuttke, and D. Streitferdt (2019). 'Interactive Content Objects as GOLDi-Lab', in *Proceedings of the The Experiment@ International Conference, exp.at* 2019, pp. 229–230.

Henry, J. (1996). 'Control laboratory teaching via the world wide web', in *Proceedings of 1996 ASEE Annual Conference Proceedings*, pp. 1.123.1–1.123.4.

Heradio, R., L. de la Torre Cubillo, D. Galan, F. J. Cabrerizo, E. Herrera-Viedma, and S. Dormido (2016a). 'Virtual and remote labs in education: A bibliometric analysis', in *Computers & Education*, Vol. 98, pp. 14–38.

Heradio, R. and S. Dormido (2016b). 'Virtual and remote labs in control education: A survey', in *Annual Reviews in Control*, Vol. 42, pp. 1–10.

Hesselink, L., D. Rizal, E. Bjornson, S. Paik, R. Batra, P. Catrysse, D. Savage, and A. Wong (2003). 'Stanford CyberLab: Internet assisted laboratories', in *International Journal of Distance Education Technologies IJDET*, Vol. 1, Nr. 1, p. 19.

Hu, M., W. Hu, and H. Zhou (2016). 'NCSLab remote experimentation for control engineering education in Wuhan University', in *Proceedings of 35th Chinese Control Conference, CCC* 2016, pp. 7416–7421.

de Jong, T., M. Linn, and Z. Zacharia (2013). 'Physical and virtual laboratories in science and engineering education', in *Science,* Vol. 340, pp. 305–308.

Kara, A., E. Ece Aydın, M. Efe Özbek, and N. Cagiltay (2010). 'Design and development of a remote and virtual environment for experimental training in Electrical and Electronics Engineering', in *Proceedings of 9th International Conference on Information Technology Based Higher Education and Training, ITHET 2010*, pp. 194–200.

Kirstein, J. (1997). 'Interaktive bildschirmexperimente — neues multimedia-labor im fachbereich physik', in *Internal report of Technical University of Berlin.*

Ko, C. C., B. M. Chen, and J. Chen (2004). *Creating Web Based Laboratories*, Springer Ed., p. 300.

Kostulski, T. and S. Murray (2010). *The National Engineering Laboratory Survey: A Review of the Delivery of Practical Laboratory Education in Australian Undergraduate Engineering Programs*. Accessible at https://trove.nla.gov.au/work/177743133?selectedversion=NBD50632027.

Lahoud, H. and T. Krichen (2010). 'Networking labs in the online environment: Indicators for success', in *The Journal of Technology Studies*, Vol. 36, Nr. 2, pp. 31–40.

Lerro, F., S. Marchisio, M. Plano, M. Protano, and O. Von Pamel (2008). 'A remote lab like an educational resource in the teaching of the physics of electronic devices', in *Proceedings of Interactive Collaborative Learning, ICL 2008*, p. 10.

López, S., A. Carpeño, and J. M. Arriaga (2015). 'Remote laboratory eLab3D: A complementary resource in engineering education', in *Revista Iberoamericana de Tecnologías del Aprendizaje, RITA*, Vol. 10, Nr. 3, pp. 160–167.

Lustig, F., P. Brom, P. Kuriščák, and J. Dvořák (2018). 'DIY Hands-on remote laboratories in physics with arduino', in *Proceedings of the 15th International Conference on Remote Engineering and Virtual Instrumentation, REV 2018*, p. 8.

Ma, J. and J. V. Nickerson (2006). 'Hands-on, simulated, and remote laboratories: A comparative literature review', in *ACM Computing Surveys*, Vol. 38, Nr. 3, pp. 1–24.

Marques, A., M. Viegas, M. Costa-Lobo, A. Fidalgo, G. Alves, J. Rocha, and I. Gustavsson (2013). 'How remote labs impact on course outcomes: Various practices using VISIR', in *IEEE Transactions on Education*, Vol. 57, Nr. 3, pp. 151–159.

Márquez-Barja, J., N. Kaminski, and L. da Silva (2018). 'Assessing the impact of user interface abstraction on online telecommunications course laboratories', in *IEEE Access*, Vol. 6, pp. 50394–50403.

McLuhan, M. (1988). '*Laws of media: The new Science*', University of Toronto Press.

Mejías, A. (2012). '*Aportaciones a los laboratorios remotos en los estudios de ingeniería. Interacción de elementos virtuales y reales mediante realidad aumentada: el laboratorio remoto aumentado*', PhD. Thesis.

Müeller, D. and H. Erbe (2007). 'Collaborative remote laboratories in engineering education: Challenges and visions', in *Using remote labs in education*, University of Deusto Ed., pp. 35–59.

National Research Council (USA) Chemical sciences roundtable (1999). *Impact of Advances in Computing and Communications Technologies on Chemical Science and Technology*, National Academies Press, p. 236.

Nedic, Z., J. Machotka, and A. Nafalski (2003). 'Remote laboratories versus virtual and real laboratories', in *Proceedings of 33rd Frontiers in Education, FIE 2003*, p. 6.

Neustock, L. T., G. Herring, and L. Hesselink (2017). 'Remote experimentation with massively scalable online laboratories', in *Proceedings of 14th Remote Engineering and Virtual Instrumentation, REV 2017*, pp. 258–265.

Nickerson, J. V., J. E. Corter, S. K. Esche, and C. Chassapis (2007). 'A model for evaluating the effectiveness of remote engineering laboratories and simulations in education', in *Computers & Education*, Vol. 49, pp. 708–725.

Orduña, P., L. Rodriguez-Gil, J. Garcia-Zubía, I. Angulo, U. Hernández-Jayo, and E. Azcuenaga (2016). 'LabsLand: A sharing economy platform to promote educational remote laboratories maintainability, sustainability and adoption', in *Proceedings 46th IEEE Frontiers in Education Conference, FIE 2016*, p. 6.

Ozvoldova, M. and F. Schauer (2016). '*Remote Laboratories: in Research-based education of real world phenomena*', P. Lang Publishers, p. 183.

Predko, M. (1999). *Programming and Customizing 8051 microcontroller, TAB Books*, p. 528.

Quintas, M. R., M. T. Restivo, J. Rodrigues, and P. Ubaldo (2013). 'Let's Use Haptics!', in *International Journal of Online and Biomedical Engineering iJOE*, Vol. 9, pp. 65–67.

Ramo, S. (1957). 'A new technique for education', in *Engineering and Science*, Vol. 21, pp. 17–22.

Restivo, T., J. C. Alves, and A. Cardoso (2011). 'Experimenta@Portugal', in *Proceedings of IEEE Global Engineering Education Conference EDUCON 2011*, pp. 1207–1210.

Restivo, T., J. Rodrigues, and F. Chouzal (2014a). 'Let's Work with AR in DC Circuits', in *Proceedings of International Conference on Interactive Collaborative Learning, ICL2014*, pp. 884–885.

Restivo, T., F. Chouzal, J. Rodrigues, P. Menezes, B. Patrão, and J. B. Lopes (2014b). 'Augmented reality in electrical fundamentals', in *International Journal of Online and Biomedical Engineering*, Vol. 10, Nr. 6, pp. 68–72.

Restivo, T., A. Cardoso, and A. Mendes (2015). *Online Experimentation: Emerging Technologies and IoT*, IFSA Publishers, p. 496.

Richter, T., D. Boehringer, and S. Jeschke (2011). 'LiLa: A european project on networked experiments, automation', in *Communication and Cybernetics in Science and Engineering 2009/2010*, Springer, pp. 307–317.

Rodrigues, J., T. Andrade, P. Abreu, and T. Restivo (2017). 'Adding augmented reality to laboratory experimentation', in *Proceedings 4th Experiment@International Conference, exp.at 2017*, pp. 135–136.

Rodríguez-Gil, L., J. Garcia-Zubía, P. Orduña, A. Villar, and D. Lopez-de-Ipiña (2019). 'New approach for conversational agent definition by non-programmers: A visual domain-specific language', in *IEEE Access*, Vol. 7, pp. 5262–5276.

Rodriguez-Gil, L., J. García-Zubía, P. Orduña, and D. Lopez-de-Ipiña (2017). 'Towards new multiplatform hybrid online laboratory models', in *IEEE Trans. On Learning Technologies*, Vol. 10, Nr. 3, pp. 318–330.

Romero, S. (2015). '*AAAS: Modelo de evaluación automática de competencias en el laboratorio remoto visir, a través de learning analytics y rúbricas de aprendizaje*', PhD Thesis.

Selmer, A., M. Goodson, R. Watson, A. Braumann, M. Kraft, S. Baguley, M. Abbott, and C. Callegari (2011). 'An undergraduate weblab using the SIMATIC PCS7 process control system', in *Using remote labs in education*, Ed. University of Deusto, pp. 375–384.

Shen, H., Z. Xu, B. Dalager, V. Kristiansen, Ø. Strøm, M. S. Shur, T. A. Fjeldly, J.Q. Lü, and T. Ytterdal (1999). 'Conducting laboratory experiments over the internet', in *IEEE Trans. On Education,* Vol. 42, Nr. 3 pp. 180–185.

Silva, J. B., K. C. N. Silva, and S. M. S. Bilessimo (2016). 'A estratégia de integração de tecnologia na educação do grupo de trabalho em experimentação remota móvel', in *Revista Tecnologias na Educação*, Anno 8, Vol. 17, pp. 1–14.

Sousa, N., G. R. Alves, and M. G. Gericota (2010). 'An integrated reusable remote laboratory to complement electronics teaching', in *IEEE Transactions in Learning Technologies*, Vol. 3, No. 3, pp. 265–271.

Soysal, O. A. (2000). 'Computer integrated experimentation in electrical engineering education over distance,' in *Proceedings of American Society for Engineering Education Annual Conference, ASEE 2000*, pp. 1.164.1–1.641.10.

de la Torre, L., L. T. Neustock, G. Herring, J. Chacon, F. García, and L. Hesselink (2020). 'Automatic generation and easy deployment of digitized laboratories', in *IEEE Trans on Industrial Informatics, accepted for publication*, DOI 10.1109/TII.2020.2977113.

Vargas, H., G. Farias, J. Sánchez, S. Dormido, and F. Esquembre (2013). 'Using augmented reality in remote laboratories', in *International Journal of Computers, Communications & Control, IJCCC*, Vol. 8, Nr. 4, pp. 619–631.

Viegas, C., A. Pavani, N. Lima, A. Marques, I. Pozzo, E. Dobboleta, V. Atencia, D. Barreto, F. Calliari, A. Fidalgo, D. Lima, G. Temporao, and G. Alves (2018). 'Impact of a remote lab on teaching practices and student learning, in *Computers & Education,* Vol. 126, pp. 201–216.

de Vries, P. R. Klaassen, D. Ceulemans, and M. Ionnides (2017). *Emerging Technologies in Engineering Education: Do we need them and can we make them work?* 4TU. Centre for Engineering Education, p. 32.

Wang, N., Q. Lan, X. Chen, G. Song, and H. Parsaei (2020). *Development of a Remote Laboratory for Engineering Education*, CRC Press, Taylor & Francis Group, p. 97.

Wisintainer, M. A. (1999). '*RExLab: Laboratório de Experimentação Remota com o microcontrolador 8051*', Master Thesis, Dir. João Bosco da Mota Alves, (https://repositorio.ufsc.br/xmlui/handle/123456789/81085).

Zappatore, M., A. Longo, and M. Bochicchio (2015). 'The bibliographic reference collection GRC2014 for the online laboratory research community', in *Proceedings of 12th Remote Engineering and Virtual Instrumentation, REV 2015*, pp. 24–31.

Zhang, P. and D. Galleta, Eds. (2015). *Human-Computer Interaction and Management Information Systems: Foundations*, Ed. Routledge, pp. 1–14.

Zutin, D. G., M. E. Auer, C. Maier, and M. Niederstätter (2010). 'Lab2go — A repository to locate educational online laboratories', in *Proceedings of IEEE Global Engineering Education Conference, EDUCON 2010*, pp. 1741–1746.

Zvacek, S. M. and T. Restivo (2018). 'Guidelines for effective online lab assignments contributions to the discussion', in *Proceedings of IEEE Global Engineering Education Conference, EDUCON 2018*, pp. 1442–1446.

Index

Lightning Source UK Ltd.
Milton Keynes UK
UKHW021837180221
378989UK00001B/43

9 781786 349422